LABORATORY HEALTH AND SAFETY HANDBOOK

LABORATORY HEALTH AND SAFETY HANDBOOK

A Guide for the Preparation of a Chemical Hygiene Plan

WITHDRAWN

R. Scott Stricoff
Vice President and Managing Director
Safety and Environmental Health
Arthur D. Little, Inc.
Cambridge, Massachusettes

Douglas B. Walters
Head, Chemical Health and Safety
National Toxicology Program
National Institute of Environmental Health Sciences
National Institutes of Health (DHHS)
Research Triangle Park, North Carolina

A WILEY-INTERSCIENCE PUBLICATION
John Wiley & Sons, Inc.
NEW YORK / CHICHESTER / BRISBANE / TORONTO / SINGAPORE

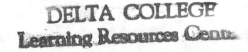

Copyright © 1990 by John Wiley & Sons, Inc.

All rights reserved. Published simultaneously in Canada.

Reproduction or translation of any part of this work
beyond that permitted by Section 107 or 108 of the
1976 United States Copyright Act without the permission
of the copyright owner is unlawful. Requests for
permission or further information should be addressed to
the Permissions Department, John Wiley & Sons, Inc.

Library of Congress Cataloging in Publication Data:
Stricoff, R. Scott.
    Laboratory health and safety handbook: a guide for the
preparation of a chemical hygiene plan/R. Scott Stricoff, Douglas
B. Walters.
        p. cm.
    Includes bibliographical references and index.
    ISBN 0-471-61756-3
    1. Chemical laboratories—Safety measures—Handbooks, manuals,
etc. I. Walters, Douglas B. II. Title.
QD51.S92 1990
542'.1'0289—dc20   90-12538
                                                            CIP

Printed in the United States of America

10 9 8 7 6 5 4 3 2 1

To
Anita, Jessica and David
Dianne and Patti

PREFACE

The field of health and safety continues to grow in complexity. As we learn more about the hazards in our workplaces, the job of managing risks to provide a safe, healthful workplace becomes increasingly challenging. Superimposed upon the inherent desire of the responsible employer to provide a safe workplace are the complex regulatory requirements of the Occupational Safety and Health Administration and the Environmental Protection Agency.

The laboratory is a workplace which presents special health and safety issues. The nature of the facility, the workers, and the tasks being performed all differ markedly from the stereotypical industrial workplace. However, the need for effective health and safety programs in the laboratory is very real, and is no less important than the need for programs in an industrial setting.

This book has been prepared to assist those responsible for laboratory health and safety in developing better programs. Particular attention has been focused on the new OSHA standard addressing hazardous materials used in laboratories. That standard is performance oriented, meaning latitude is given to laboratories in complying; however, the programs which are required may be new to many laboratory facilities.

The book includes a specific discussion of the "Chemical Hygiene Plan" required by OSHA's new laboratory standard. In addition, a chapter on the recognition and anticipation of laboratory hazards is included. Also a section of several chapters addressing other "management system" elements of the safety program, e.g., operating procedures, documentation, emergency planning, and training is present.

Another major section of the book covers facilities and design of laboratories. General laboratory design and the special needs of containment facilities are reviewed. Fire safety, ventilation, emergency power, eyewash and emergency shower facilities, and human factors in design are each discussed in a separate chapter.

Personal exposure evaluation and protection issues are covered in the third major section of the book. Topics included are chemical exposure assessment, medical surveillance, personal protective equipment, and respirators. Advice is provided on programs appropriate for hazard control in laboratories.

Several special hazards are discussed in another separate section of the book. There are chapters on chemical incompatabilities, biohazards, radiation, controlled substances, and waste management.

Finally, there is a brief review of major regulations affecting laboratory health and safety programs. A bibliography is also provided to guide the reader to additional information on the topics covered in the book.

This book will be particularly useful to the laboratory manager, supervisor, or administrator who is responsible for health and safety, but is not a health and safety professional. The book provides practical information and guidance with which an effective program consistent with OSHA requirements can be developed and implemented.

March 1990
Cambridge, Massachusettes   R. SCOTT STRICOFF
Research Triangle Park, North Carolina   DOUGLAS B. WALTERS

CONTENTS

Chapter 4

Hygiene Plan   33

Chapter 5

Standard Operating Procedures   41

Chapter 6

Documentation 47

Chapter 7

Emergency Plans 51

Chapter 8

General Laboratory Design 59

Chapter 9

Barrier System Design 69

Chapter 10

Construction Materials   89

Chapter 11

Fire and Explosion Protection   93

Chapter 12

Local Exhaust Ventilation   111

Chapter 13

Emergency Power   143

Chapter 17

Medical Surveillance   189

Chapter 18

Protective Equipment   197

Chapter 19

Respirators   209

Chapter 20

Chemical Incompatibility   225

Chapter 21

Biohazards   239

Chapter 22

Radiation   251

Chapter 23

Controlled Substances   271

Chapter 24

Waste Management 281

Chapter 25

Regulations 303

CHAPTER 1

RECOGNITION AND ANTICIPATION OF LABORATORY HAZARDS

I INTRODUCTION

The initial step in the development of an overall laboratory health and safety program is to identify the chemical, biological, and physical hazards that are present in the facility. This chapter provides a framework for developing a component of the program and shows how to recognize hazards initially and then as components of the working program. Continual use of preliminary surveys, hazardous chemical inventories, material safety data sheets (MSDSs), health and safety packages, and annual baseline surveys provide a facility's health and safety officer (HSO) with a methodology for ongoing recognition and anticipation of laboratory hazards. Information in the chapters that follow will help a facility build a hazard evaluation and control program that is based on the proper and timely recognition of hazards.

II RECOGNITION AND ANTICIPATION OF HAZARDS: METHODOLOGY

One of the primary tasks of a comprehensive health and safety program is to identify, evaluate, and recognize potential employee exposures to all toxic chemicals.

TABLE 1−1   Development of a Laboratory Safety and Health Program

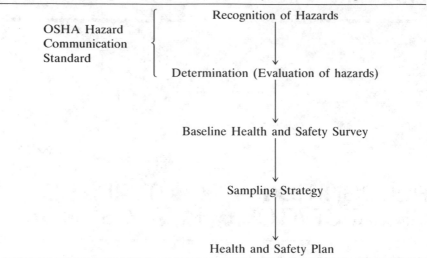

Material safety data sheets for all chemicals in the laboratory are useful in the initial recognition of chemical hazards in the workplace. Based on the information they provide, the HSO can perform a preliminary survey to ensure that the proper controls are in place for the safe handling of the chemicals. The MSDSs also assist the HSO in conducting a hazardous chemical inventory and hazard assessment. The HSO can then identify potential chemical, biological, and physical agent exposures and conduct a baseline survey to quantify these exposures. Finally, the HSO can develop a health and safety plan based on the results of the surveys.

This chapter outlines a recommended approach to recognizing potential health hazards, utilizing the MSDSs and baseline surveys. A large portion of these programs is based on the requirements found in the Hazard Communication Standard (29 CFR 1910.1200) of the Occupational Safety and Health Administration (OSHA). Table 1−1 gives the sequence for a generic hazardous chemicals recognition program; the major topics are discussed in the remainder of the chapter.

A   Material Safety Data Sheets (MSDS)

Each facility must ensure that vendors supply MSDSs for the chemicals it purchases. A vendor should make sure that the facility receives the MSDS before, or along with, the first shipment of a newly purchased chemical product.

If a vendor delivers a shipment of a new or reformulated chemical without an MSDS, the facility should contact the chemical manufacturer, importer, or distributor to obtain one as soon as possible. The OSHA Hazard Communication Standard not only requires manufacturers to supply MSDSs, but also obligates purchasers to actively pursue MSDSs that they need but have not received.

Suggested Methods

A number of systems can aid accomplishment of this task.

- At a minimum, a facility should check orders of vendor chemicals to confirm that an MSDS is on file. One approach to ensure that current MSDS are maintained is to conduct a physical inspection of each shipment, or review packing slips or invoices and then cross-check the MSDSs on file. This system suggests some relationship between the MSDS coordinator and the purchasing or receiving official.

- Alternatively, a facility may give its purchasing department complete responsibility for ensuring that current MSDSs have been received from the vendor and have been properly distributed. Purchase orders and supply contracts should include requirements that a current MSDS be supplied as required by the standard.

- In a more sophisticated approach, a facility could use a computer to track vendor products, MSDSs, and shipment orders. Some computer systems are capable of automatically cross-referencing products that require a new vendor MSDS, as well as purchase records, requirements, and hardware/software sources.

If a facility does not receive an MSDS with the first shipment of a new or reformulated chemical, it should request the vendor to supply one. The facility should make such requests in writing and keep copies on file. Figure 1–1 presents a sample MSDS request letter.

A facility must check all MSDSs for completeness. The vendor should not leave any block on the MSDS sheet blank. At the very least, the vendor should indicate "not applicable" (NA) or "no information available" (ND or No Data) with respect to each category not addressed.

Each facility is responsible for maintaining a file of current MSDSs covering materials used in each work area, and for seeing that these documents are readily available to employees in their work areas during work hours. In addition, the facility must advise its employees of the location of this information and provide access to it.

There are several technical reference sources and computer data bases

Mr. John Vendor, Purchasing Manager
XYZ Company
P.O. Box 374
Anywhere, USA

Dear Sir:

On October 8, 1986, VA Facility X received a shipment of hydrochloric acid, shipment 374, manufactured by XYZ Company. We would like to call the following situation(s) to your attention:

Situation 1 — We did not receive a Material Safety Data Sheet (MSDS) with this product shipment and do not currently have an MSDS product for this on file. Please provide us with a current MSDS, or suitable product safety bulletin as soon as possible.

Situation 2 — The MSDS we received with this product shipment was incomplete in the categories noted below:

Note deficiency

Please provide us with a complete MSDS as soon as possible.

It is company policy to request an MSDS for products we purchase from our vendors on delivery, as required by the OSHA Hazard Communication Standard. If you cannot provide us with a completed MSDS, or if you have any questions pertaining to this request, please contact me at (#) or in writing.

We would appreciate your timely response to our request.

Sincerely,

John Doe
MSDS Coordinator

VA Facility X

FIG. 1–1:   Sample vendor letter.

that a facility can use to check the accuracy of its MSDS information (see Section III. A).

   Suggested ways by which a facility can help ensure that MSDSs are readily accessible to employees and made available to interested parties include the following:

- A facility should provide employee access to MSDSs, so that the workers will be privy to a detailed source of hazards. A facility should provide employees with MSDSs for their review *before* they use any chemical(s) that are so documented. In many cases the MSDS should be utilized to prepare a safety protocol, which employees should review before use of the chemical. An MSDS alone may not provide enough information, particularly facility-specific information, crucial to the safe use of the chemical. Also, a facility should train its employees how to access the MSDS files. A facility may post instructions in the work area, or it may pass this information on to its employees verbally, by memorandum, or by newsletter. A facility should keep MSDSs in notebooks, files, data bases, or in any other convenient medium that remains in, or close to, the immediate work area.

- Alternatively a facility may provide its employees with MSDSs in documentary form (written operating procedures, manuals, product bulletins, etc.), as long as the material is readily accessible and provides the required information. This approach may be useful in addressing the hazards of an operation rather than *individual chemicals*. However, MSDSs must be available for each chemical used in an operation.

When reading an MSDS, one should consider the following points:

1. An MSDS often outlines only minimum precautions for the safe handling of a chemical, to document fire and explosion hazards, spill or leak procedures, special protection information, and special precautions. Thus, if a facility's policy, or the reader's own judgment, suggests that more stringent procedures are in order, such additional measures should be instituted.

2. In many cases, a facility will write emergency first aid procedures and handling precautions to deal with a worst-case scenario, such as an extensive exposure to a hazardous chemical. With this knowledge, an employee might conclude, for a seemingly minor incident, that he or she needs no immediate medical attention, even though the MSDS might require it. In the absence of an informed opinion from a designated health professional, however, a prudent response is the best policy. Follow the MSDS first aid procedure.

3. When an MSDS fails to mention a particular detrimental health effect, it should not be assumed that the substance is hazard-free. The

vendor may not have been privy to test results when the MSDS was prepared.

4. An MSDS should have no blank spaces. When a facility receives an incomplete MSDS, an agent should question the supplier. As indicated previously, if no data are available, the MSDS should so state in the space provided.

B Preliminary Health and Safety Survey

The purpose of a preliminary walk-through survey is twofold: to identify potential health hazards and to collect relevant operational, environmental, and personnel background information for the future assessment of those health hazards. At a minimum, the surveyor(s) must collect the following information in a preliminary walk-through survey.

1. A list of potentially hazardous materials (i.e., a hazardous material inventory), including toxic chemicals and biological and physical agents that may cause harm to a facility's employees, as defined in the OSHA Hazard Communication Rule (29 CFR 1910.1200). The minimum information required for each chemical agent consists of material name, frequency of use, and amount of use:

2. Descriptions of operations and tasks involved in the handling, storage, and disposal of hazardous materials and their by-products; also, descriptions of operations that use hazardous physical agents or stressors.

The following information and factors are required for the initial determination of potential health hazards:

1. Type of operation and specific operational characteristics that may influence the presence and/or the degree of a hazardous exposure, that is,

- narrative of operation including documented written procedure, if available;
- type of equipment used;
- physical/chemical factors of the operation; for example, temperature, pressure, chemical reactions; and
- specific task performed by employees.

2. Frequency and duration of the operation and potential exposure of employees to a hazard.

3. Number of personnel, both male and female.

4. Description of control measures (e.g., type of personal protection equipment, engineering controls, such as specific type of laboratory hoods and shielding, and administrative controls, such as training and rotation of personnel.

5. Rudiments of each operation. For example, are chemicals being heated with or without combustion? Are dry materials being dumped or mixed? Are liquids being sprayed? How often is each operation run?

6. Obvious signs of exposure, including:

- airborne dust, smoke, or mist;
- accumulations of dust, liquid, or oil on machines, on the floor, and/ or the ledges;
- odors from solvent vapors or gases;
- unusual taste;
- burning or throat/nose irritation;
- hazardous operations being performed during unsupervised times or, special tasks being performed by maintenance personnel.
- presence of procedures for responding to emergencies, such as chemical spills, leaks, explosions, and fires;
- complaints of employees about such symptoms as
  skin rash or dermatitis; coughs, tightness of the chest, difficulty in breathing; stuffy noses and persistent colds;
  headaches, dizziness, or light-headedness;
  loss of appetite, fatigue, nausea; and
  numbness in the fingers, hands, arms, or legs;
- persistence of symptoms, or improvement when people are away from work;
- provision of special medical tests, such as blood and urine tests, lung function tests, and x-rays, for employees on special jobs;
- high turnover rate on certain jobs; and
- conduct of sampling or industrial hygiene surveys and their results.

The facility's health and safety officer should collect the background information listed above, with the assistance of other laboratory personnel. The HSO should also address biological and radiological hazards. Before generating a list of potential health hazards for each operation, the HSO should evaluate the background information.

Table 1−2 presents (an example of a preliminary survey checklist that HSOs often use when making laboratory industrial hygiene surveys) incorporates the physical and health hazard information necessary to evaluate

TABLE 1−2 Program for Hazard Recognition Program

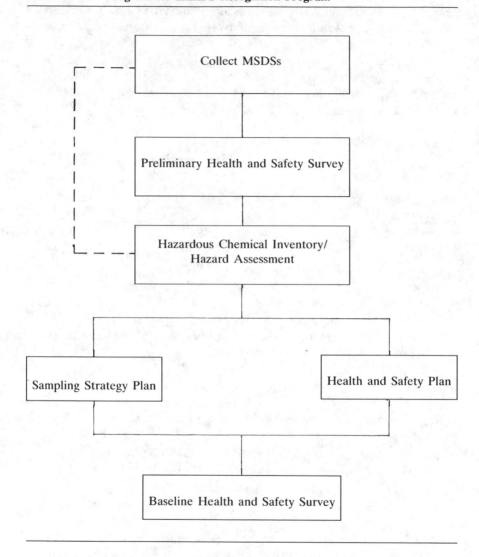

an operation, according to the OSHA Hazard Communication Rule and good industrial hygiene practice.

Table 1−3 is an example of hazardous agents found in operational areas of toxicology laboratories. Additional agents may also be present in the areas, and the checklist should include them as well.

TABLE 1−3 Physical Constraints/Hazardous Agents in Toxicology Laboratories

Operation	Physical Constraints/ Agent/Exposure
Receiving	Lifting, twisting Solvents Test chemical Positive controls
Storage	Lifting Solvents Positive controls
Dose preparation	Test chemical Lifting, repetitive motion Noise (mixing) Radiation (trace) Solvents (e.g., alcohol, DMSO) Biological agents Positive controls
Dose administration	Lifting, repetitive motion Dilute test chemical Radiation Biological agent
Histology	Hydrochloric acid Picric acid 27−40% Formaldehyde 95% Alcohol 10% Formalin xylene Paraffin Eosin stain Glacial acetic acid Mercuric oxide Hematoxylin Basic fuchsin Metanil yellow Trypan blue stain Geimsa stain Repetitive motion, eye strain

TABLE 1−3 (Continued)

Analytical chemistry	Solvents
	Reagents
	Repetitive motion
	Radiation (ionizing, nonionizing)
Cagewash	Aerosols (bedding)
	Biological agents
	Test chemical
	Positive controls
	Noise
	Heat stress
	Lifting, twisting
Boiler plant	Sodium hydroxide
	Phosphates
	Sulfuric acid
	Chlorine
	Heat stress
	Lifting, twisting
	Asbestos (insulation)
	Noise
Engineering/maintenance	Welding:
	Ultraviolet radiation
	Oxides of nitrogen
	Fluorides
	Flux fumes
	Ozone
	Solvents
	Glues
	Pesticides
	Noise
	Cleaning fluids/disinfectants
	Test chemical
	Positive controls
	Lifting, twisting
	Asbestos (insulation)

Development of a Hazardous Chemical Inventory/Hazard Assessment

As stated in the OSHA Hazard Communication Rule:

> Effective May 25, 1986, facilities need to have, as a part of their written hazard communication program, a list of the hazardous chemicals known to be present using an identity (chemical name) that is referenced on the appropriate MSDS. The list of hazardous chemicals can be compiled for the workplace as a whole or for individual work areas. The list of hazardous chemicals does not need to include: hazardous wastes; wood or wood products; "articles" as defined by OSHA; and food, drugs or cosmetics intended for personal consumption by employees in the workplace.

The information outlined above becomes useful as workplace lists are compiled. Suggested methods for developing and maintaining a workplace/work area hazardous chemical inventory listing follow.

As required by the federal Hazard Communication Rule [29 CFR 1910.1200(e)], a facility is responsible for developing a list of hazardous chemicals to be included as a part of its written hazard communication program.

One approach a facility might take in compiling a list of all its workplace chemicals is to conduct a preliminary survey of each work area and/or review its purchasing and inventory records (as discussed above). Based on these surveys, the facility should compile the complete workplace chemical listing. Each inventory listing should be kept in its respective work area. A sample of a hazardous chemical inventory form is provided as Figure 1–2.

Once a facility has completed its chemical inventory, it should cross-reference its contents with MSDSs received to date. When an MSDS states that a material is not hazardous, and a check reveals that the chemical is indeed not listed as hazardous under the OSHA Hazardous Communication Rule, the facility may consider dropping the chemical from its list. A laboratory may also discover that chemicals identified in the inventory lack MSDS. In such cases, the laboratory should take one of the following steps:

- Ask the HSO for advice on whether the chemical is hazardous and whether, consequently, it would require an MSDS and need to be included on the list.
- Send for the MSDS to determine whether the manufacturer is required to provide one, and if the chemical is hazardous.

CHEMICAL NAME_____ STORED AT_____
MATERIAL/INVENTORY CONTROL NUMBER___ SUPPLIER_____
is used in this department.
FROM: DEPARTMENT_____
BY: NAME_____ DATE_____
Complete one sheet for each material and for each supplier and send it to the
facility HazCom Coordinator.

CHARACTERISTICS OF MATERIALS TO LOOK FOR:

Gas	Liquid	Solid
Pressurized	Poisonous	Corrosive
Toxic	Carcinogen	Flammable
Combustible	Nuclear	Alkaline
Oxidizer	Pyrophoric	Infectious solvent
"Reactive hazard"	"Housekeeping supply"	Very dusty
Heavy mist	Fume problem	Strong odor

Incompatible with_____
Sensitive to: heat   light   shock   pressure
****

Frequency of Use: _____
Amount of Use: _____
Total Quantity: _____

FIG. 1-2:   Material Inventory Questionnaire.

A facility must keep the workplace or work area list(s) of hazardous
chemicals up to date to satisfy the requirements of the OSHA Hazard
Communication Standard. In addition, it must see that the list is available
for employees in the workplace or work areas.

A facility may use a number of methods in *maintaining* the workplace
hazardous chemical list. Such methods may vary in sophistication and
complexity, based on facility size, procurement procedures, and the number
of chemicals required in the workplace. Regardless of the method employed, the facility should update the list of hazardous chemicals by
adding new chemicals each time they are received at the facility or
created by an operation change, and deleting them from the chemical list
when no longer present. The facility should also include the list as part of
its written hazard communication program. A facility with centralized
chemical purchasing procedures may find it convenient to cross-reference
chemical orders directly with the facility master list and make additions as
needed. A facility with decentralized chemical purchasing responsibilities
in each work area should provide a separate new material notice for each
new chemical. A sample form is provided as Figure 1-3. A facility

TO: Hazardous Materials Coordinator _____
       Department/Supervisor _____
FROM: Purchasing Department_____; Engineering_____;
       Research_____; Receiving_____; Other_____

Name of Person ordering chemical _____
This is to inform you that:
       _____a material is being deleted from our inventory.
       _____a new or different material has been requested for (purchase) (is in
       use) by this department.

Material _____   Supplier _____
Purchase Order No. _____
Brand Name or Chemical Name _____
Quantity_____   Frequency of Use_____   Amount to be Used_____

Characteristics of this material that suggest this notification: (circle one or more
that apply):
       gas, liquid, solid, poisonous, corrosive, toxic, carcinogenic, acid base,
       flammable, combustible, nuclear, osicizer, pyrophoric, incompatible
       with _____
       sensitive   to: heat,   light,   shock,   pressure,   cold,   water
       reactivity hazard of _____, infectious hazard

Some known information:
       U.N. No. _____   DOT No. _____
       EPA No. _____   CAS No. _____
A Material Safety Data Sheet (MSDS) has (not) been requested from the supplier
and is (not) in the permanent MSDS file at _____.

Date MSDS is needed:_____/_____/_____

FIG. 1−3: New Material Notice

employee should fill out the form whenever new chemicals are purchased.
The facility may also use the form to notify its hazardous materials
coordinator of the deletion of a material from inventory, or the introduction
of a new material.

When a contractor brings chemicals on site, the facility should obtain a
list of these materials to supplement the workplace inventory until the
chemicals are removed from the facility by the contractor.

Finally, some states have enacted right-to-know laws and have prepared
lists of hazardous chemicals which facilities should consult in their attempts
to comply with state requirements for listing hazardous chemicals.

Briefly, the OSHA Hazard Communication Rule states that laboratory
personnel are responsible for hazard determinations and MSDSs. To
meet this requirement, laboratory personnel must:

- evaluate the physical and health hazards of chemicals produced or imported;
- identify and consider the available scientific evidence concerning such hazards;
- follow mandatory Appendices A and B of the standard in conducting hazard determinations;
- accept that materials listed in 29 CFR Part 1910, Subpart Z, or in the threshold limit values (TLVs) for chemical substances and physical agents in the work environment of the American Conference of Governmental Industrial Hygienists (ACGIH) are hazardous;
- accept that chemicals listed as carcinogenic by the National Toxicology Program (NTP), or the International Agency for Research on Cancer (IARC), or regulated by OSHA as carcinogenic are, in fact, carcinogenic for hazard communication purposes, as stated in the OSHA Hazardous Communication Rule;
- assume that mixtures that are untested, on the whole, have the same hazards as their hazardous components present at 1% or more (0.1% for carcinogens);
- see that hazard determination procedures are documented in writing;
- see that MSDSs are developed or obtained for each hazardous chemical;
- see that MSDSs are documented (in English) and contain the information required by the standard;
- see that MSDSs accurately reflect the scientific data considered in the hazard determination;
- see that copies of MSDSs for each hazardous chemical in the workplace are maintained in the workplace, and ensure that this material is readily available to employees during each workshift when they are in their work area(s); and
- see that employees are properly trained and understand the specifics of the MSDSs.

III HAZARD ASSESSMENT

OSHA has provided guidance for making the hazard assessments that the HSO or hazardous materials coordinator must perform upon receipt by a facility of an MSDS. The OSHA standard lists approximately 600 hazardous chemicals. Further designations of a hazardous chemical are listed below.

A Suggested Methods

Information Sources. A facility should use all available sources of health and physical information to assess the potential hazards that a chemical may possess. These sources include, but are not limited to:

1. the hazard determination criteria given in Appendix B and the recommended literature references listed in Appendix C of the OSHA Hazard Communication Standard;
2. information on the physical – chemical properties, toxicology data, and human health effects, as well as environmental fate and effects data;
3. all adequate health and safety information obtained from the supplier;
4. primary resources used for hazard evaluation:
   (a) National Library of Medicine
       (i) On-line computerized medical literature analysis and retrieval system (MEDLARS)
       (ii) Immediate access to more than 2.5 million references on more than 500,000 chemicals
   (b) Dialog
       (i) More than 200 data bases
       (ii) More than 90 million records
       (iii) A wide variety of subject matter, including chemistry, medicine, and biosciences, as well as science technology data
   (c) Lexis/Nexis: federal and state regulatory/legal data.

Determination of a Physical Hazard A *chemical* is considered to constitute a *physical hazard* if there is scientifically valid evidence that it is a combustible liquid, a compressed gas, an explosive, flammable, an organic peroxide, an oxidizer, pyrophoric, unstable (reactive), or water-reactive, as defined by the OSHA Hazard Communication Standard [Section (c) — Definitions].

Some definitions given by OSHA that are useful in physical hazard determination are listed here.

"Combustible liquid" means any liquid having a flash point at or above 100°F (37.8°C), but below 200°F (93.3°C), or higher, the total volume of which makes up 99% or more, of the total volume of the mixture.

"Compressed gas" means:

1. a gas, or mixture of gases, having, in a container, an absolute pressure that exceeds 40 psi at 70°F (21.1°C); or
2. a gas, or mixture of gases, having, in a container, an absolute

pressure that exceeds 104 psi at 130°F (54.4°C), regardless of the pressure at 70°F (21.1°C);or

3. a liquid having a vapor pressure that exceeds 40 psi at 100°F (37.8°C), as determined by the American Society for Testing and Materials (ASTM-D-323−72).

"Explosive" means a chemical that causes a sudden, almost instantaneous release of pressure, gas, and heat, when subjected to sudden shock, pressure, or high temperature.

"Flammable" means a chemical that falls into one of the following categories:

1. "Aerosol, flammable" means an aerosol that, when tested by the method described in 16 CFR 1500.45, yields a flame projection exceeding 18 in. at full-valve opening, or a flashback (a flame extending back to the valve) at any degree of valve opening:

2. "Gas, flammable" means a gas that, at ambient temperature and pressure, forms a flammable mixture with air at a concentration of 13% by volume or less; or a gas that, at ambient temperature and pressure, forms a range of flammable mixtures with air wider than 12% by volume, regardless of the lower limit.

3. "Liquid, flammable" means any liquid having a flash point below 100°F (37.8°C), except any mixture having components with flash points of 100°F (37.8°C) or higher, the total of which makes up 99% or more of the total volume of the mixture.

4. "Solid, flammable" means a solid, other than a blasting agent or explosive, as defined in 29 CFR 1910.109(a), that is liable to cause fire through friction, absorption of moisture, spontaneous chemical changes, or retained heat from manufacturing or processing, or that can be ignited readily and, when ignited, burns so vigorously and persistently as to create a serious hazard. A chemical shall be considered to be a flammable solid if, when tested by the method described in 16 CFR 1500.44, it ignites and burns with a self-sustained flame at a rate greater than 0.1 in./s along its major axis.

"Organic peroxide" means an organic compound that contains the bivalent −O−O− structure and may be considered to be a structural derivative of hydrogen peroxide where one or both of the hydrogen atoms has been replaced by an organic radical group.

"Oxidizer" designates a chemical other than a blasting agent, or explosive, as defined in 29 CFR 1910.109(a), that initiates or promotes combustion in other materials, thereby causing fire either of itself or through the release of oxygen or other gases.

"Pyrophoric" designates a chemical that ignites spontaneously in air at a temperature of 130°F (54.4°C) or below.

"Unstable (reactive)" designates a chemical that, in the pure state, or as produced or transported, vigorously polymerizes, decomposes, condenses, or becomes self-reactive under conditions of shock, pressure, or temperature.

"Water-reactive" designates a chemical that reacts with water to release a gas that is either flammable or presents a health hazard.

B Health Hazard Determination

1. To determine whether a chemical is a *health hazard*, a facility should evaluate the data to establish:
   (a) whether there is statistically significant evidence, based on at least one study conducted in accordance with scientific principles, that acute or chronic health effects may occur in employees exposed to it (refer to the full definition of health hazard as provided in the OSHA standard);
   (b) whether it meets any of the health hazard criteria, as defined in the OSHA standard, that would make it a carcinogen: a corrosive, highly toxic irritant; a sensitizer; a toxic material; or an agent that could harm any organ and/or system of the body, human or animal;
   (c) whether it is listed in OSHA Regulation 29 CFR 1910 Subpart Z, or in the ACGIH threshold limit values for chemical substances and physical agents in the work environment;
   (d) whether it is listed in the IARC monographs or the (NTP), Annual Report on Carcinogens, or is regulated as a carcinogen by OSHA (29 CFR 1910 Subpart Z); if so, this will be considered as conclusive evidence of its carcinogenicity; and
   (e) whether the results of any studies that were designed and conducted according to established scientific principles indicate statistically significant evidence that the chemical poses a potential health hazard.
2. To assess that a *health hazard* exists for a mixture, a facility should evaluate to determine:
   (a) whether the mixture has been tested as a whole: if so, the health hazard determined and the ingredients (if known) that contributed to this condition are indicated on the MSDS;

(b) whether the mixture has *not been tested* as a whole for health hazard: in this case the health hazard determination for component chemicals is used in the evaluation of any components of the mixture that comprise 1% or greater ($\geq$ 1% for carcinogens); and

(c) whether it is determined that a component (< 1%, or < 0.1% for a potential carcinogens) present in the mixture could be released in atmospheric concentrations that would exceed the established OSHA or ACGIH exposure limit for the component, or would present a health hazard at these concentrations, then the mixture would be considered hazardous.

A facility or workplace may contain common consumer products, such as household detergents and cleansers, soap, and type correction fluid (e.g., White-Out®). These items may be excluded from the hazard assessment, *provided* they are used in the same manner and in the same approximate quantities as would be expected in their typical consumer applications. If, for example, a commercial sodium hypochlorite solution (e.g., Clorox®) is used regularly to disinfect work surfaces for microbial contamination, it should be included in the program.

IV SAMPLING STRATEGY PLAN

Chapter 16 discusses the sampling strategy plan in detail. The plan, developed from the preliminary survey and hazardous chemical inventory for the facility, must provide a good technical and management approach for the industrial hygiene monitoring requirements at the laboratory. If potential exposure to a hazardous chemical exists, the facility must offer a rational explanation as to why sampling, engineering controls, or personal protective equipment was or was not provided. The facility's HSO is responsible for providing clearly written documentation and keeping it in the files.

V BASELINE HEALTH AND SAFETY SURVEY

The baseline health and safety survey constitutes implementation of the sampling strategy plan. The baseline survey should include the quantitative monitoring of agents, the evaluation of personal protection equipment, and the evaluation and verification of the effectiveness of engineering controls. In addition to these technical issues, the survey should address the management of industrial hygiene programs. Chapter 4 discusses these evaluation techniques and control programs in detail.

The HSO is responsible for conducting the baseline survey and documenting the results. The preliminary survey checklist can be completed with the results of the baseline survey and used as the baseline survey report. This document can be updated as necessary; a similar survey would provide adequate documentation of the industrial hygiene survey.

Whenever operations that involve use of hazardous chemicals in a facility are modified, the industrial hygiene survey process should be fully implemented and documented as necessary.

CHAPTER 2

RESPONSIBILITIES

I INTRODUCTION

Development, implementation, and maintenance of a comprehensive health and safety program require the participation of many different people. This chapter identifies key laboratory personnel on whom the responsibility for the program rests, and defines the nature of their responsibilities.

II RESPONSIBILITIES

A Laboratory Manager

The primary responsibility of the laboratory manager is to implement the health and safety program. The laboratory manager is required to:

(a) ensure that all work is conducted in accordance with the chemical hygiene plan and the location's health and safety program;

(b) prepare procedures for dealing with accidents that may result in the unexpected exposure of personnel, or the environment, to a toxic substance;

(c) select the appropriate control practices for handling hazardous substances;

(d) report to the safety officer the location of work areas in which toxic substances and potential carcinogens will be used, and ensure that the inventory of hazardous substances is properly maintained; and

(e) prepare a safety plan for use of hazardous substances.

B Health and Safety Officer

The primary responsibilities of the health and safety officer (HSO) are to monitor all aspects of safety and to serve as primary source for health and safety information. The HSO is required to:

(a) assist the laboratory manager in defining hazardous operations, designating safe practices, and selecting protective equipment;

(b) make copies of the approved safety plan available to the technical and support staff;

(c) obtain, review, and approve standard operating procedures, detailing all aspects of proposed laboratory activities that involve hazardous agents;

(d) ensure that all personnel obtain medical examinations and the protective equipment necessary for the safe performance of their jobs;

(e) ensure that technical and support staff receive instructions and training in safe work practices and in procedures for dealing with accidents involving test substances;

(f) monitor the safety performance of the staff to ensure that the required safety practices and techniques are being employed;

(g) conduct formal laboratory inspections quarterly to ensure compliance with existing laboratory policies and government regulations;

(h) arrange for workplace air samples, wipe samples, or other tests to determine the amount and nature of airborne and/or surface contamination, and use data to aid in the evaluation and maintenance of appropriate laboratory conditions;

(i) assist the radiation safety officer when necessary;

(j) document and maintain compliance with all local, state, and federal regulatory requirements;

(k) develop rules and procedures for safe practices, assist in the development and review of health and safety plans; consult, advise, and make recommendations on all health and safety matters;

(l) develop health and safety training plans and programs, conduct training courses, and establish safety references;

(m) report to the laboratory manager incidents (1) that cause personnel to be seriously exposed to hazardous chemicals or materials, such as through the inoculation of a chemical through cutaneous penetration, ingestion of a chemical, or probable inhalation of a chemical, or (2) that constitute a danger of environmental contamination;

(n) investigate accidents and report them to the laboratory manager;

(o) investigate and report in writing to the laboratory manager any significant problems pertaining to the operation and implementation of control practices, equipment, or the facility;

(p) dispose of unwanted and/or hazardous chemicals and materials; and

(q) ensure that action is taken to correct work practices and conditions that may result in the release of toxic chemicals.

C Employees

Employees are required to:

(a) gain in understanding of, and act in accordance with, the safety requirements established by the laboratory;

(b) wear the safety equipment and personal protective equipment necessary to perform each task they are assigned; and

(c) report to the laboratory manager or the health and safety officer all facts pertaining to every accident that results in exposure to toxic chemicals, and any action or condition that may exist that could result in an accident.

III CHEMICAL HYGIENE PLAN CONSIDERATIONS

The chemical hygiene plan is required to specify in detail the responsibilities for the health and safety program. Responsibilities for program design, implementation, and review should be covered. These responsibilities should be consistent with the assignment of other management responsibilities in the laboratory.

CHAPTER 3

TRAINING

I INTRODUCTION

Laboratory employees may encounter hazards of various types, including biological, chemical, and radioactive, as well as fire in a laboratory. A laboratory should familiarize its employees with these hazards and the risks they involved. Individuals properly trained in handling hazardous materials are better equipped to minimize the risk of exposure to themselves, their peers, and the environment. A comprehensive training program will provide proper orientation in the use of safety equipment and the implementation of related procedures and policies. However, the success of a training program depends on management's support of these programs and the utilization by the trained employee of the information developed.

II RESPONSIBILITY FOR TRAINING

The ultimate responsibility for ensuring a safe working environment rests with the employee. He or she should assume an active role in maintaining a safe working environment by reporting any problems, or noncompliance with policies, to the supervisor or laboratory manager. All employees are accountable to their peers and, therefore, should fully utilize the information provided during formal and informal training sessions. Any staff member who does not understand a policy or procedure should consult the health and safety officer (HSO) for clarification.

The HSO should make sure that all employees are aware of the hazards in the workplace and the safety measures available, and that they know how to use the safety equipment properly. Laboratory personnel qualified in the areas of health and safety — namely, the HSO, the laboratory manager, or a supervisor — should develop and implement training programs. The laboratory should provide these programs to all new employees and conduct them regularly. The laboratory should also properly document all training sessions, placing records of those attending and subject matter covered in the facility's health and safety files.

Laboratory management should encourage proper training and attitudes toward safety among its employees. As stated in the *NIH Guidelines for the Laboratory Use of Chemical Carcinogens of the National Institutes of Health*.

> Employees should be provided with sufficient information to understand the potential hazards that can affect them personally. Employees should be periodically advised about (1) the possible sources of exposure, (2) adverse health effects associated with exposure, (3) laboratory practices and engineering controls in use and being planned to limit exposure, (4) the use and purpose of any recommended environmental and medical monitoring procedures, and (5) their responsibilities for following proper laboratory practices to help protect their health and provide for the safety of themselves and fellow employees.

In brief:

- Employees should utilize the information provided, comply with federal, state, and local regulations, and report problems to the laboratory manager.
- The HSO, laboratory manager, and/or supervisor should communicate the potential hazards associated with the work plan, encourage proper training and attitudes toward safety, make sure that all employees are aware of the safety measures available and how to properly use them, provide training on a regular basis, and properly document all training sessions and list those who attended.

III TRAINING PROGRAM COMPONENTS

Each laboratory should offer a comprehensive employee training program. Such a program should provide training to new employees before work is assigned to them, with additional training continued throughout their employment. Specific phases of the program should be repeated at least annually, and records should be maintained by the laboratory to indicate whenever an employee has completed a specific training session.

A Training Requirements for the Health and Safety Officer

Each laboratory should have a qualified health and safety officer. The level of qualification and time commitment of the HSO will, of course, depend on the size of the facility and the complexity of its work. For a laboratory conducting moderate to large-scale work with hazardous materials (e.g., an experimental toxicology laboratory doing *in vivo* work, or a large chemical or pharmaceutical research location), qualifications should include the following:

- Bachelor's degree, with a major in chemistry, biology, chemical engineering, or a closely related field.
- At least 2 years' experience (part-time) in occupational health and ˉsafety *along with* completion of courses in general occupational health and hazard control indicating the acquisition of successively greater levels of knowledge regarding industrial hygiene. Training should be refreshed, with additional training at an interval not exceeding 18 months. The health and safety officer may have other responsibilities within the organization; however, the amount of time devoted explicitly to health and safety should be commensurate with the scale of the laboratory operations. (A master's degree in industrial hygiene or a bachelor's degree in industrial hygiene with one year or experience is an acceptable substitute for this experience requirement).

The HSO is responsible for implementing the laboratory's various training programs.

B Training Requirements for All Employees

Personnel who are potentially exposed to hazards at the laboratory should be provided with written materials on the nature of the hazards given a formal training program. This training should be conducted by a qualified safety person and properly documented; it should provide instruction in handling radiolabelled materials, where applicable.

All personnel should also receive basic training in fire safety, including

- hazard awareness,
- proper techniques for handling and storing flammable liquids,

- briefing on the alarm system and emergency evacuation preplanning, and
- hands-on training in the use of fire extinguishers.

Before using any laboratory hood, personnel should be trained in its proper use and monitoring procedures. The operator's education and training should include a daily visual and smoke tube inspection.

Each laboratory must implement, a respirator program that meets the requirements of the Occupational Safety and Health Administration (OSHA: 29 CFR 1910.134) and includes a training session in the proper use and limitations of respirators. This training should provide all employees with an opportunity to handle the respirators discussed, to have one fitted properly and tested for fit, and to wear the unit in a normal atmosphere for an extended period at time. The program should also include a discussion of engineering controls, respirator selection, the potential health hazard when a respirator is not used, and recognition and handling of emergencies. The laboratory should designate the person who is to be responsible for each program element.

C Recommended Outline of Training Program

The following training courses/components should be performed for laboratory personnel. Depending on the nature of the work done at the laboratory, some of the sessions listed may not be necessary.

- Respirator protection and fit-testing program:

  - selection, use, and maintenance of respirators,
  - compliance with OSHA 1910.134,
  - fit-testing procedures, and
  - emergency-use respirators.

- Handling hazardous chemical — acquisition to disposal: receiving, transportation within the facility, storage, handling, and disposal.
- fire training — prevention and response
- Emergency response and evacuation
- Handling radiolabeled materials
- Interpretation of a material safety data sheet (MSDS)
- First aid and CPR
- Engineering controls

- General laboratory safety
- Personal hygiene
- Protective clothing

IV   OSHA HAZARD COMMUNICATION REGULATION

OSHA adopted a Hazard Communication Regulation (29 CFR 1910.1200: see Chapter 25, Regulations) to ensure that employees would be properly informed of the chemical hazards associated with the use of any hazardous materials they handled in the workplace. A comprehensive hazard communication program is to become the vehicle to accomplish this information transfer. It should include container labeling and other forms of warnings, MSDS, and employee training. This standard, as it applies to laboratories training their employees, states:

> 1910.1200 (b) (3) (iii)
> Employers shall ensure that laboratory employees are apprised of the hazards of the chemicals in their workplaces in accordance with paragraph (h) of this section.
>
> . . .
>
> 1910.1200 (h) Employee Information and Training
>
> Employers shall provide employees with information and training on hazardous chemicals in their work areas at the time of their initial assignment, and whenever a new hazard is introduced into their work area.
>
> (2) Training
> (i) Methods and observation that may be used to detect the presence or release of a hazardous chemical in the work area.
> (ii) The physical and health hazards of those chemicals.
> (iii) The measures employees can take to protect themselves from these hazards, such as specific work practices, emergency procedures, and personal protective equipment.
> (iv) The details of the hazard communication program.

As noted previously, many states have "right-to-know" laws, which predate the OSHA regulations. For the most part, federal law has superseded them. However, in some instances, portions of the state laws are still intact. Each laboratory should contact the state to determine whether its hazard communication regulations, if any, are obsolete or still apply.

V OSHA'S VOLUNTARY TRAINING GUIDELINES

The OSHA Hazard Communication Standard was promulgated in 1984. Previously, OSHA had developed and distributed Voluntary Training Guidelines (published in the *Federal Register*: 49 FR 30290), which were to serve as a blueprint for training requirements in its standards (e.g., hazard communication). OSHA has consolidated these guidelines into seven steps, as shown in Figure 3–1. These guidelines may be useful to laboratories in their training program planning.

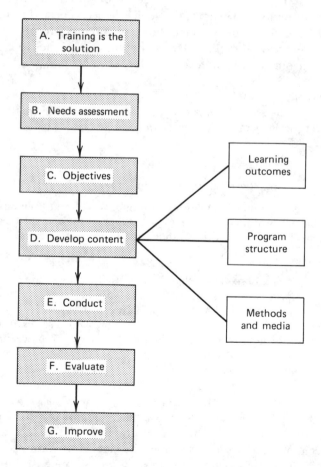

FIG. 3–1: Summary of OSHA's voluntary training programs.
SOURCE: Occupational Health and Safety, May 1987.

VI CHEMICAL HYGIENE PLAN CONSIDERATIONS

The chemical hygiene plan must address health and safety training provided for laboratory workers. The frequency and content of the training must be specified and should be consistent with the program described previously. The hygiene plan should also indicate who the trainer(s) will be, and qualifications of the trainer(s).

VII TRAINING RESOURCES

There are many sources of health and safety training materials and programs. Thus the listing that follows is not exhaustive; however, it indicates the range of materials available from various sources of audiovisual training aids.

National Audiovisual Center
National Archives and Records Service
General Services Administration
Reference Section CF
Washington, DC 20409
(1−301−763−1896)

- Principles of Physical and Chemical Containment − Unit III
- Certification of Class II (Laminar Flow) Biological Safety Cabinets
- Effective Use of the Laminar Flow Biological Safety Cabinets
- Nobody's Perfect
- Selecting a Biological Safety Cabinet
- Hazard Control in the Animal Laboratory

3M Hazard Awareness Program
3M Center
St. Paul, MN 55144

- Fire and Extinguishers
- Laboratory Hoods

Office of Continuing Education Harvard School of Public Health

- Safety, Health and Ventilation Issues in the Laboratory
- Certification of Biological Safety Cabinets

- Fundamentals of Industrial Hygiene

American Chemical Society
1155 16th Street, N.W.
Washington, DC 20036

- Chemical Carcinogens

Fisher Scientific Company
711 Forbes Avenue
Pittsburg, PA 15219

- 28 Grams of Prevention

Other Sources
BNA films
BNA Communications Inc.
9439 Kew West Ave.
Rockville, MD 20850
NIOSH ERC's
National Institute of Occupational Safety and Health (NIOSH)
Educational Resource Centers (ERC's)
4676 Columbia Pkwy
Cincinnati, OH 45226–1998
AIHA and NIOSH sponsored courses
American Industrial Hygiene Association (AIHA)
175 Wolf hedges Parkway
Akron, OH 44311–1087

CHAPTER 4

HYGIENE PLAN

I INTRODUCTION

Effective management of occupational health and safety at laboratories requires the preparation of a comprehensive chemical hygiene plan. Once written, such a plan is a valuable resource and reference for all persons protected by its provisions. In addition, the development of the plan is, in and of itself, a useful exercise in hazard identification, evaluation, and control.

Developing an accurate and up-to-date plan offers certain benefits. It requires each laboratory to carefully define what hazards exist in the facility, what programs are necessary to control the hazards, and what means can be used for assessing its effectiveness and degree of implementation. In addition, a hygiene plan demonstrates a positive and preventive approach to safety and health, and it reflects a laboratory's commitment to safe practices and operation. Finally, it provides a certain amount of uniformity of work practices within and between laboratories. In its rule, "Occupational Exposure to Hazardous Chemicals in Laboratories," the Occupational Safety and Health Administration (OSHA) has indicated that hygiene plans will be required for all laboratories.

This chapter provides an overview of the information necessary to prepare a hygiene plan that will satisfy OSHA's requirements. The remainder of the book discusses in greater detail the issues that should be addressed in laboratory safety programs.

II OSHA LABORATORY STANDARD

A Scope

The scope of the OSHA standard is limited to "all employers engaged in the laboratory use of toxic substances . . . .".

First, OSHA defines "laboratory use of toxic substances" as the handling or use of such substances in which all of the following conditions are met: (1) chemical manipulations are carried out on a "laboratory scale"; (2) multiple chemical procedures and/or chemicals are in use (3) the procedures involved are not part of a production process, nor in any way simulate a production process; and (4) protective laboratory practices, which may include the use of appropriate equipment, are available and in common use to minimize the potential for employee overexposure to hazardous chemicals.

The term "laboratory scale" is defined as work with substances in which the containers used for reactions, transfers, and other handling of substances are designed to be easily and safely manipulated by one person. "Laboratory scale" excludes workplaces that function to produce materials in commercial quantities.

Workplaces or activities that do not satisfy the foregoing definitions are not considered to be laboratories, and they will continue to be regulated under existing OSHA standards.

B Chemical Hygiene Plan

OSHA has required that each employer whose activities fall within the definitions discussed above establish a chemical hygiene plan for protecting employees from health hazards inherent in the toxic substances used in that laboratory. However, OSHA expects that the provisions developed for the substances that fall within the narrowly defined scope will apply to other hazardous materials in the facility as well. The preamble states that the impact of the standard is potentially broad, since most laboratories handle at least one substance that is either a toxic substance or a carcinogen as these terms were defined. Therefore, such laboratories would be required to implement work practices that would serve as effective protection against substances not explicitly covered by the standard that may be nevertheless potentially hazardous. Consequently, this standard, with its emphasis on safe handling and use of toxic substances, lends itself to protection against hazards that go beyond the toxic substance definition, with no cost attendant to the added protection.

The content of the hygiene plan must include the following items:

- Standard operating procedures (SOPs) to be followed when laboratory work involves the use of hazardous chemicals.
- Criteria to determine the need for, and the nature of, the exposure control strategies to reduce personnel exposures. These strategies include engineering and administrative controls and the use of personal protective equipment.
- A requirement that control measures including lab hoods and other local exhaust ventilation, be properly selected, designed, installed, and maintained, along with procedures to ensure satisfaction of the requirement.
- Information and training procedures.
- A provision for medical consultation and evaluation.
- Circumstances under which a particular laboratory operation will require approval prior to implementation.
- Identification of personnel responsible for implementation and maintenance of the hygiene plan.
- For work performed with carcinogenic materials and other particularly hazardous chemicals, additional protective measures.

III HOW TO COMPLY WITH OSHA REQUIREMENTS

The OSHA-defined chemical hygiene plan is limited in its scope to the hazardous chemical worker exposure issues currently addressed by 29 CFR 1910, Subpart Z. However, most laboratories will find it more practical and realistic to address health and safety through a comprehensive program and will find no advantage to the creation of a hygiene plan separate from the overall laboratory health and safety plan. Rather, the OSHA requirements for a hygiene plan will be met in most instances through an upgrading of the overall health and safety plan to include OSHA-required elements.

A Statement of Policy

A clear, concise statement of policy is the cornerstone on which a health and safety program should be built. If the employees of a laboratory can see and read the commitment that top management has made to their safety through safe operating practices, the first step to safety vigilance has been taken. Conversely, employees could construe the absence of a firm and forcefully written commitment by senior management as a lack of concern, a mindset that could lead to a relaxed and careless attitude on the part of both investigators and employees. It must be stressed, however, that management should support and reaffirm its statement of policy by consistent actions that demonstrate the kind of resolve represented in the written statement.

The statement of policy should incorporate the following characteristics:

- It should be specific to the location.
- It should clearly set out the safety and health goals and the commitment embraced by senior management.
- It should state that all employees are responsible for excellent safety and health performance.
- It should identify the technical resources and staff members to whom employees can turn for assistance in fulfilling their safety responsibilities.
- It should be signed by the person or persons for whom it purports to speak.

As various changes that warrant updating the safety and health policy statement occur, the effect of these changes should be reflected promptly in the policy statement. Newly revised policy statements should be sent to all holders of the location's health and safety manual, and posted in conspicuous areas in the laboratory.

B Standard Operating Procedures (SOPs)

Chapter 5 describes the nature and content of SOPs to be included in the hygiene plan. The objective of these procedures is to represent the "how to" part of the safety plan for particular activities and investigations within the laboratory. The OSHA standard explicitly states the need for SOPs.

C Description of Control Strategies

OSHA recognizes the need for effective exposure control strategies. The various engineering and administrative control strategies, as well as the use of personal protective equipment, are addressed elsewhere in this manual. All are summarized briefly below.

1 Engineering Controls

Engineering controls form the first line of defense against exposures to toxic or hazardous agents for laboratory personnel. The extensive volume of air-moving and ventilation equipment present in many laboratories constitute the primary engineering control factor at these facilities.

OSHA requires that the hygiene plan describe the locations in which, as well as the conditions under which, engineering controls must be identified and described.

The final aspect of engineering controls is perhaps the most important. The effectiveness and efficiency of a ventilation system, whether it is a local exhaust hood or a negative-pressure barrier facility, depend on regular preventive maintenance and routine measurements. Appropriate sections of the hygiene plan should outline the procedures for performing the maintenance and taking the measurements.

2 Administrative Controls

Many laboratories rely heavily on the proper administrative oversight of laboratory hazards. As described in other sections, the programs for training, access, documentation, project review, labeling and posting, as well as other administrative activities, should be included in the hygiene plan.

3 Personal Protective Equipment (PPE)

Most of the laboratories affected by the OSHA standard use some personal protective equipment, including limited use of respiratory protection.

D Training

The hygiene plan must describe the training provided to employees that is consistent with the requirements of Chapter 3. This training requires the accumulation of pertinent health and safety information about all hazards present in the laboratory, followed by communication of that information to employees who might be affected. In addition, the hygiene plan should provide clear and careful documentation of the content, scheduling, and attendance at training sessions.

E Recordkeeping

In addition to the recordkeeping required for the training programs, other types of recordkeeping, as described in Chapter 5, are required.

These records serve as a basis on which compliance with a laboratory's health and safety program can be determined and assessed.

F Emergency Response

Special kinds of emergency are both laboratory-specific and generic for all laboratory activity. The emergency response portion of the hygiene plan should include appropriate instructions and procedures to be followed in the event of natural disasters, fires and explosions, and chemical leaks or spills. The specifics of these instructions and procedures are described throughout the rest of this manual.

G Periodic Review of Hygiene Plan

The hygiene plan, of necessity, is specific to the time, place, and conditions that exist during its development. As conditions, activities, or personnel change, the laboratory must amend and update the hygiene plan to keep it current.

Each plan should be drafted to ensure that any major change at the facility will trigger a review of the hygiene plan. In addition, the laboratory should review the plan periodically, even if no specific changes have occurred, to assure that the plan is current and incorporates recent technical advances.

H Response to Overexposure

The rule adopted by OSHA addresses the technical exposure evaluation and the medical surveillance activities that are triggered by either an actual overexposure or a reasonable belief on the part of an employee that he or she has been subjected to an overexposure. Specifically, OSHA requires an exposure evaluation for employees who, as a consequence of a laboratory operation, procedure or activity, reasonably suspect or believe that they have sustained an overexposure to a toxic substance."

"Exposure evaluation" is defined as "an assessment of the conditions present during a laboratory operation, procedure, or activity for the purpose of determining if an employee has, or may have been, overexposed to a toxic substance."

In essence, OSHA expects a laboratory to be able to respond to an unforeseen accident or incident in a manner that permits either qualitative or quantitative assessment of the risk of exposure. Appropriate measures could include the use of direct-reading instrumentation or colorimetric detector tubes, or calculations to estimate the concentration of airborne toxic substances. The hygiene plan, though, should identify what will be done — and who will do it — in response to an alleged overexposure.

The provision of medical attention in response to a perceived over-exposure is a two-step process in the OSHA rule. First of all, a medical consultation must be made available at no cost to the employee. Second, if suggested by this consultation, the employee must be given access to appropriate diagnostic or surveillance procedures.

CHAPTER 5

STANDARD OPERATING PROCEDURES

I INTRODUCTION

A comprehensive health and safety program should include documents that provide detailed descriptions of standard methods or operations used within the facility. These documents, commonly referred to as standard operating procedures (SOPs), should be used by all employees. They should describe how to perform the method or operation in question. Also, SOPs should be written clearly and precisely, so that an individual responsible for a particular procedure, or piece of equipment, can easily understand them.

II RESPONSIBILITIES

A Management

Management should delegate responsibility for the preparation of SOP documents to a qualified staff member or organizational element responsible and knowledgeable in the areas of interest. Management is responsible for determining the adequacy of the SOPs prepared. Also, management should request from all personnel a signed statement that they have read and understand all SOPs pertinent to their duties. Finally, management should provide for the placement of these records in a permanent archive.

B Employees

Safety is an ongoing and active concern in a laboratory. All employees are responsible to their peers for ensuring that specific standard safety procedures are followed by everyone involved. Safety in the laboratory depends on critical judgments and how each team member applies the information that is made available and observes the procedures. The approval of the laboratory manager and/or the safety officer should be required before any deviations from these procedures are permitted. Each employee should read all SOPs associated with his or her particular area of responsibility and should acknowledge that the SOPs are understood.

III DEVELOPMENT OF THE PROCEDURE

A good standard operating procedure is one that is clearly stated and realistic in scope. A laboratory should prepare SOPs for all its routine, repetitive, and unique operations as well as for nonroutine events (e.g., spills). The laboratory should take care to develop required and ancillary procedures in a consistent manner. The following general information, in conjunction with stepwise procedures, may be useful in establishing an accurate document:

- photographs, graphs, or illustrations,
- flowcharts, if applicable,
- appropriate forms,
- equipment used during a particular operation (a cross-reference to the equipment SOP should be made),
- reagents used,
- persons responsible in the event of an emergency, and
- publishedliterature used as a supplement.

The format of all SOPs should be consistent. An official form should be prepared, incorporating:

- facility name;
- subject;

- division, department, section affected by or using the procedure;
- issue date of the original document or current revision;
- an indication that the revision replaces an earlier procedure;
- signature or initials of both the responsible issuing individual and the individual approving the procedure on behalf of management;
- a number on each page with reference made to the total number of pages occupied by a given procedure; and
- an identification code for each SOP.

Figure 5−1 shows a typical format for presenting a standard operating procedure.

Cell Biology/Microbiology Unit

STANDARD OPERATING PROCEDURE # CB/M−100B

SOP Creation

Date Issued	Page 1 of 1	Written by:

APPROVALS:	Q.A. Coordinator:	Sr. Professional: Unit Manager:

FIG. 5−1: Typical SOP Format.

IV GOOD LABORATORY PRACTICES: OTHER GOVERNMENT AGENCIES

For laboratories that conduct nonclinical studies for research or marketing permits for products regulated by the Food and Drug Administration, requirements for the preparation of standard operating procedures are outlined in Title 21 of the Code of Federal Regulations (CFR).

For laboratories that conduct studies on health effects, environmental effects, and chemical fate testing, 40 CFR, Part 792, Subpart E, prescribes good laboratory practices relating to standard operating procedures.

The U.S. Environmental Protection Agency (EPA) also has established good laboratory practice standards for laboratories conducting studies that support applications for pesticide products regulated by the EPA. For pertinent standard operating procedures for these facilities, see 40 CFR, Part 160, Subpart E.

V CHEMICAL HYGIENE PLAN CONSIDERATIONS

The chemical hygiene plan should include a listing of the SOPs relevant to the health and safety program. Copies of the actual SOPs most directly related to the safety program may be included as appendices to the hygiene plan.

A laboratory may choose to develop SOPs related to health and safety in the following areas:

- visitors' access to testing areas;
- entry and exit from the limited access areas;
- employee training;
- medical surveillance;
- respirator protection and fit-testing programs;
- eye protection;
- personal protective equipment;
- general housekeeping practices;
- carcinogen handling;
- ventilation system maintenance;
- storage and transportation of hazardous materials;
- spill cleanup, accident, and emergency response;
- waste disposal;

- use of radiolabeled materials, infectious agents, and/or controlled substances; and
- general safety policies including (but not limited to) use of safety glasses; mouth pipetting; dry sweeping; eating, drinking, smoking, applying cosmetics, and chewing gum; actions to be taken in case of fire and/or explosion; personnel assignments; and evacuation routes and notification procedures.

RESOURCES

Hoover, B. K. J. K. Baldwin, A. F. Velnar, C. E. Whitmire, C. L. Davies, and D. W. and Bristol, *Managing Conduct and Data Quality of Toxicology Studies.* Princeton, NJ: Princeton Scientific Publishing 1986.

CHAPTER 6

DOCUMENTATION

I INTRODUCTION

This chapter addresses documentation required to support a comprehensive health and safety program.

Over the past several years, increases in regulatory activity, environmental health and safety litigation, and employee awareness of potential hazards in the workplace have forced employers to maintain accurate records on health and safety programs. To assure that this documentation program is in place, and is accurate and effective, the following items must be accomplished:

- development of documentation procedures (standardized format, sign-off, etc.),
- distribution of documents — both external and internal, and
- maintenance of documentation program.

Given the regulatory and legal environment, documentation programs go well beyond maintaining the proper papers in files. Each laboratory must establish a paper trail that will provide employee, employer, and regulatory agencies with an accurate representation of how hazardous agents are handled in the facility.

A   Needs Assessment of Documentation Program

The development of a documentation program begins with a review of the pertinent health and safety regulations on documentation requirements.

B   Development of Documentation Process

A laboratory should develop a standardized format for each type of document, agreed to by all the personnel who will use the form. For example, the worksheet shown in Chapter 12 (Fig. 12−9) would be useful for documenting of a hood monitoring program. Laboratory personnel who perform tests and use the forms to document them should sign and date these documents. Also, the user, the health and safety officer, the principal investigator, and representatives of laboratory management should review each written program for health and safety plans, respiratory protection provisions, protective clothing, and health and safety training. Also, the laboratory should maintain a sign-off sheet as part of the file material kept on each written program.

C   Distribution

The laboratory is responsible for distributing, posting, and/or circulating the document to its employees, to management, and to state or federal agencies. The laboratory must also make the documents available for agency inspections. A distribution system should include listings of where the documents are and who has signed them, statements that employees have read the documents, and notification that individuals or groups have received their copies.

D   Documentation Program Maintenance

Modification of written programs, updates of new monitoring techniques, and physical storage of files are key components of a documentation program. Clearly, an automated system with built-in notification for document updates would be advantageous. The use of microfiche for long-term storage of records would also be beneficial.

Retention time for these documents is also an issue. Local, state and federal regulations require a time span from 30 years to indefinite storage of records. Plans for storage should reflect consideration of ways to accommodate any loss of records caused by physical damage (weather, etc.) and access limited to personnel with need-to-know status.

II CHEMICAL HYGIENE PLAN CONSIDERATIONS

The chemical hygiene plan should tell who will maintain the documentation, describe such documentation, and state the retention policy. Examples of the types of documentation that may be maintained are:

- health and safety officer's credentials;
- health and safety plan;
- health and safety training (including sign-off statement from project personnel on reading and understanding SOPs and health and safety plan);
- archival information, such as employee roster, medical surveillance logs, biological monitoring results, air sampling results, ventilation system performance and maintenance records, waste disposal records, respirator fit-testing program, and accident and incident reports;
- SOPs;
- serious accident/incident records; and
- record of exit/entry to restricted-access areas.

CHAPTER 7

EMERGENCY PLANS

I INTRODUCTION

It is especially important that all employees of a laboratory understand and be well trained in the procedures they must follow in the event of a fire, explosion, toxic chemical release, or accident. Every laboratory should have a comprehensive emergency plan ready for implementation during such events. No employee should have to "second guess" the first priority (i.e., to save one's self, or save the experiment). Nor should employees have to hesitate in deciding if and when they can safely shed their personal protective equipment (PPE) to avoid all chance of exposure to a dangerous chemical.

II EMERGENCY RESPONSE

The procedures recommended in the event that a chemical carcinogen or toxic material is spilled are as follows:

1. Leave the room immediately and, if any chemical has contacted your eyes or skin, wash thoroughly using the eyewash and/or shower, as necessary. Discard any protective clothing that may have been contaminated.

51

2. Notify personnel in the immediate area to evacuate and post the area with warning signs.

3. When a person is contaminated, it is necessary to notify for assistance the Poison Control Center and/or laboratory management.

4. Before returning to the spill area, an employee should be made aware of the neutralization process for the specific compound spilled. Persons not wearing the appropriate personal protective equipment and clothing should be restricted from areas of spills or leaks until cleanup has been completed.

5. The minimum personal protective equipment that should be worn during cleanup consists of:

(a) respirator, with the appropriate cartridge or self-contained breathing apparatus,
(b) disposable Tyvek® jumpsuit,
(c) disposable Tyvek shoe protectors and Tyvek head cover,
(d) safety glasses, and
(e) disposable gloves (at least two dissimilar pairs, with the inner pair taped to the sleeve cuff of a Tyvek jumpsuit).

6. No one is to enter the spill area alone. Only when accompanied by another appropriately dressed individual may a person enter the spill area. If no one else is available, a person desiring to enter is to remain immediately outside the room, dressed in protective equipment and ready to provide assistance in case of problems.

7. The following procedures should be observed in the cleanup of solid chemical spills:

(a) Cover the solid material with wet paper towels (using water or appropriate solvent). Avoid spreading the compound as much as possible. DO NOT DRY SWEEP.
(b) If the material is flammable, remove all sources of ignition.
(c) Ventilate the area of the spill.
(d) Carefully pick up the bulk of the material in one of the scoops provided with the spill kit.
(e) Again, with wet paper towels, wipe up any small traces of material still present.
(f) Follow the neutralization process appropriate for the spilled material to decontaminate the area.
(g) Dispose of the residue according to the recommended hazardous waste disposal procedure.

8. The following procedure should be observed in cleaning up, liquid chemical spills:

(a) Ventilate the area of the spill.
(b) If the material is flammable, remove all sources of ignition.
(c) Surround the area with an absorbent material (Solusorb, paper towels, sodium bicarbonate, sand, or vermiculite). Solusorb is available in spill kits sold commercially.
(d) Carefully spread more absorbent material onto the chemical, and try to avoid creating aerosols. Allow enough time to soak up the liquid.
(e) Carefully pick up the bulk of the material in one of the scoops provided with the spill kits.

III MEANS OF EGRESS

In planning for a timely and efficient evacuation in case of an emergency, a means of egress must be considered and planned. The principal reference, which has been widely adopted by many states and municipalities, is National Fire Protection Association Standard 101, "Code for Safety to Life from Fire in Buildings and Structures." It addresses such topics as type and number of exits, measurement of travel distance to exits, and emergency lighting.

IV ALARM SYSTEMS

An integral part of an emergency response plan is a reliable, well-designed alarm system. Principal components of such a system include:

- a control unit;
- initiating device circuiting, with appropriate connections (to manual pull-box stations, detectors, water flow alarms, etc.);
- indicating device circuiting, with appropriate connections (to bells, speakers, and off-premises alarms, etc.); and
- primary and secondary power supplies; there is great variety of systems available, ranging from local alarms to central station alarms systems.

When considering an alarm system for a specific site, details on the design, installation, inspection, and maintenance of the different types of system can be found in the following documents of the National Fire Protection Association (NFPA):

- NFPA Standard 71, "Installation, Maintenance and Use of Central Station Signaling Systems"
- NFPA Standard 72A, "Standard for the Installation, Maintenance and Use of Local Protective Signaling Systems for Guard's Tour, Fire Alarm and Supervisory Service"
- NFPA Standard 72B, "Standard for the Installation, Maintenance and Use of Auxiliary Protective Signaling Systems for Fire Alarm Service"
- NFPA Standard 72C, "Standard for the Installation, Maintenance and Use of Remote Station Protective Signaling Systems"
- NFPA Standard 72D, "Standard for the Installation, Maintenance and Use of Proprietary Protective Signaling Systems"
- NFPA Standard 72E, "Standard on Automatic Fire Detectors"
- NFPA Standard 72F, "Standard for the Installation, Maintenance, and Use of Emergency Voice/Alarm Communication Systems"
- NFPA Standard 72G, "Standard for the Installation, Maintenance and Use of Notification Appliances for Protective Signaling Systems"
- NFPA Standard 72H, "Guide for Testing Procedures for Local Auxiliary, Remote Station and Proprietary Protective Signaling Systems"

V FIRST AID

First aid recommendations supplied with chemical specific handling documents (e.g., material safety data sheets) typically assume that a physician, ambulance, or emergency medical service is available within 5–15 minutes. They reflect an out-of-hospital situation and are thus designed to cover only the initial period while awaiting professional help. First aid assistance should be provided only by trained individuals.

The following general guidelines are typically used:

1. If the victim is convulsing or unconscious, DO NOT INDUCE VOMITING. Inducing vomiting in an unconscious person is likely to aspirate the chemical into the lungs, thus spreading it and causing other complications. Ensure that the victim's airway is open and lay the victim on his or her side with the head lower than the body. IMMEDIATELY transport the victim to a hospital.

2. If the chemical ingested is an irritant, corrosive, or volatile substance, DO NOT INDUCE VOMITING. Inducing vomiting with chemicals of these types is likely to aspirate the chemical into the lungs and may harm

or destroy other tissues in the throat or mouth, thus spreading it and causing other complications. Usually the best advice is to dilute the chemical by having the victim drink 1 or 2 glasses of water until the care of a physician or paramedic has been obtained. If the chemical is very toxic, the victim may be advised to drink a slurry of activated charcoal to adsorb the chemical while awaiting medical help.

3. If the chemical ingested is not an irritant, corrosive, or volatile but is very toxic (i.e., the quantity sufficient to induce death is about 1 teaspoon or less), then, because of the high toxicity of the chemical, consider undertaking the risk of inducing vomiting. Ipecac syrup or salt water may be used in such an emergency.

4. If the chemical ingested is not an irritant, corrosive, or volatile and is low in toxicity (this covers the majority of organic and inorganic compounds), give 1 or 2 glasses of water to dilute the chemical while awaiting medical help.

5. If the chemical ingested is a concentrated acid, give the victim several glasses of very cold water to dilute it. This is because of the heat of dilution, which is released when concentrated acids are diluted.

6. If the chemical ingested is a dilute acid, give the victim several glasses of cold water and also Maalox®, milk of Magnesia or aluminum hydroxide gel to neutralize it. Avoid all carbonated beverages, since these will release carbon dioxide in the stomach.

7. If the chemical ingested is either concentrated or dilute base, give the victim several glasses of cold water to dilute it.

8. If the chemical ingested is a known or suspected carcinogen, determine from a physician whether long-term monitoring is recommended. The specific compound, exposure route, and exposure level will determine the physician's recommendation.

9. If the chemical is spilled on the skin, immediately wash it off with soap and water. If it is not very toxic, corrosive, or an irritant, and is not readily absorbed by the skin, contact a physician if any symptoms such as a redness or spots develop. If the chemical is corrosive or an irritant, in which case symptoms such as redness or spots are likely to develop quickly, contact a physician. If the chemical is toxic and readily absorbed by the skin [e.g., if there is a "skin" notation in the threshold limit values (TLVs) of the American Conference of Governmental Industrial Hygienists (ACGIH)], contact a physician immediately anyway, because there may not be any discernible symptoms on the skin.

10. If the chemical is splashed in the eyes, flush the eyes immediately for about 20 minutes. If the chemical is not an irritant, not corrosive, and not very toxic, contact a physician if any symptoms such as redness or

irritation of the eyes occur. If the chemical is an irritant, corrosive, or toxic, contact a physician immediately; again, symptoms are likely to develop quickly in such cases.

11. If the chemical is inhaled, see that the victim leaves the area immediately and breathes fresh air. If the chemical is not an irritant, not corrosive, and not very toxic, contact a physician if any symptoms such as coughing or shortness of breath occur. If the chemical is an irritant, corrosive, or toxic, contact a physician immediately — symptoms are likely to develop quickly.

Supplies to keep on hand for emergencies include:

- Ipecac syrup or table salt for inducing vomiting;
- activated charcoal for making a slurry to drink;
- Maalox®, milk of magnesia, or aluminum gel to neutralize dilute acids; and
- appropriate respirators.

Eyewash stations to flush eyes (see recommendations for eyewash stations) also should be available.

VI CHEMICAL HYGIENE PLAN CONSIDERATIONS

All laboratories should have certain information regarding emergency response, first aid, and spill cleanup readily available for use by their employees. All employees should be aware of these procedures and should understand their importance. The hygiene plan should convey this information, which includes the following:

- A standard operating procedure for accidents and emergency response, as well as an SOP to follow in the event of a spill or leak, with an explanation of the necessary steps for a spill cleanup. The format SOP should address the storage and use of emergency protective equipment. The SOP for emergency response should specify personnel assignments, evacuation routes, and notification procedures. If necessary, individual SOPs should be prepared on each specific area of emergency response.
- Emergency power generators are required on all laboratory hoods, storage units, and general ventilation systems, and are to be activated in the event of a power failure.

- Training is to be provided on emergency evacuation procedures, as well as on the alarm systems used throughout the facility. Other training programs in first aid, the use and storage of emergency protective equipment, and the use of fire extinguishers are also recommended.
- Emergency protective equipment is not to be stored in any area where test or control chemicals are stored or handled.
- The telephone number of the Poison Control Center and all other emergency numbers are to be posted in each laboratory.

CHAPTER 8

GENERAL LABORATORY DESIGN

I INTRODUCTION

Recently, experts in laboratory design have focused their attention on the proper design of high hazard or special-function laboratories that are used solely to handle highly toxic chemical and biological agents. However, new regulatory actions from OSHA and increased employee awareness have necessitated health and safety reviews of general laboratories where such chemicals and agents are handled.

II HOW TO DESIGN GENERAL-USE LABORATORIES

Figure 8–1 presents a design methodology. Designers can modify this scheme for laboratories of different types, based on the operations and personnel who will be utilizing the space. Therefore, the first step in the design process is to perform a needs assessment to gain an understanding of both user and project activities. Once this has been accomplished, the design team can define the functional requirements of the laboratory, which in turn permits the application of the functional requirements, the worker protection plan, and risk assessment to the design criteria.

In designing a general laboratory, the risk assessment should be a qualitative process review to identify any hazardous operations that should

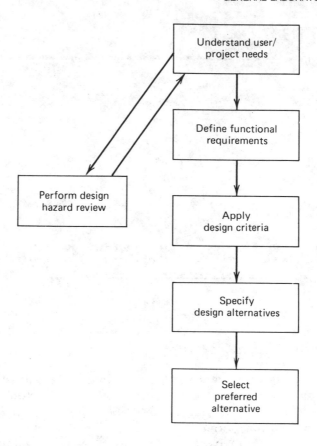

FIG. 8-1:   Laboratory design process.

not be performed in a general-use laboratory. Once the design criteria have been agreed on, the designers should execute several alternative designs before arriving at the final choice. This design methodology is usually performed by architect–engineers working closely with those who will be using the laboratory. Worker protection and risk assessments must be defined by health and safety professionals who are familiar with the evaluation and control practices for handling chemical, biological, and radiological agents, as well as fire protection and the life safety code.

The initial step in any proven laboratory design methodology is to consider the personnel who will be using the laboratory, their needs and equipment, and most important, their future project requirements. Interviews with laboratory personnel at all levels are vital to the completion of this step. Detailed lists of their activities and the chemicals they will use are helpful (Table 8–1). At a minimum, the lists should include major operations that will be conducted and the equipment and chemicals

TABLE 8–1   **Sample Laboratory Operations and Chemicals Used**

Laboratory	Examples of Various Operations	Chemicals Handled
Histology	Preparation of some solutions and stains	Hydrochloric acid Picric acid 27–40% Formaldehyde Sodium acid phosphate monohydrate 95% Alcohol
	General preparation of tissue	10% Formalin Graded alcohols Xylene Paraffin
	Slide staining: hematoxylineosin stain	Absolute alcohol Hematoxylin Mercuric oxide Hydrochloric acid Glacial acetic acid
Pathology	Preparation of some solutions and stains	See above
	General preparation of tissue	See above
	Slide staining: Gridley fungus stain	4% Chromic acid Basic fuchsin Sodium metabisulfite Hydrochloric acid 65% Alcohol Paraldehyde Metanil yellow Glacial acetic acid 95% Alcohol Absolute alcohol Xylene
Cytology	Staining slides	Various cell-specific stains e.g., Gram stains, Trypan blue stain, Geisma stain

to be used. For example, a histology laboratory would require the following facilities and equipment characteristics:

- location adjacent to solvent storage;
- Ventilation for tissue preparation and formalin dispensing;
- negative pressure;
- sinks, eyewash stations, and shower facilities; and
- access to the necropsy laboratory.

The information presented in the text and in Table 8−1 leads to the generation of a needs list (Table 8−2). In addition to the needs shown in Table 8−2, the activities under consideration may require special facilities. For example, many histology and necropsy laboratories are constructed with pass-through hoods between the necropsy and tissue trimming locations to facilitate transfer of the animals and tissues between laboratories.

When the personnel, operations, and chemicals have been identified and understood, designers can address the assessment of risk to personnel and worker protection. This type of review is known as a design hazard review (DHR). For general laboratories, a DHR should be performed for the laboratory as a whole and for specific projects if they add additional hazards to the laboratory and cannot be handled by existing engineering and personnel protective equipment controls. After the initial DHR, laboratory management should initiate a subsequent DHR only if, in their judgment, hazards associated with any new project(s) warrant it.

The following information belongs in a DHR:

- description of the process,
- list of raw materials and products,
- size and type of equipment, and
- statement of potential hazards.

TABLE 8−2   Example of a Laboratory Needs List

Secure storage space for alcohols
Protection against exhaust air reentrainment
Laboratory work stations for tissue trimming, slide preparation and staining, etc.
Storage for stains in powder form
Vented enclosure for automated tissue processing and staining equipment
Vented enclosure for weighing materials (e.g., stains)
Vented enclosure for dispensing formalin

While performing the DHR, one or more health and safety professionals will review the process for any points of uncontrolled chemical, biological, and/or radiological exposure to laboratory personnel. Designers must review each exposure point to assure that equipment controls are adequate to eliminate the exposure points. This review must cover the process from the receipt to the disposal of the hazardous agent. Figure 8–2 is an example of an entire DHR process performed by a large chemical company for its laboratory facilities.

Laboratory designers can construct a series of design criteria, based on the design review and user/needs assessment. An excellent source for design specifications is *Guidelines for Laboratory Design* [1]. A comprehensive listing of important elements of design for general laboratories can be developed from this source. As shown in Table 8–3, the designer must address all these considerations in the DHR and needs assessment. Design criteria are discussed in Reference 2, and examples are given below.

- *Worker protection.* The laboratory must minimize the risk of worker exposures. Thus, ventilation systems must be adequate in number and operational characteristics, there must be suitable changing and storage areas, eyewash and emergency showers must be available, etc. The design must also consider the risk of reentrainment of contaminated air by air handling equipment serving either the lab itself or other nearby space.

- *Traffic flow.* The lab must be designed with consideration for the patterns of movement of people and material within the laboratory. Minimization of the transport of hazardous materials from its point of receipt to storage, and from storage to working area is preferred. The use of passthrough to allow the delivery of material and dispensing of prepared dilute samples can be desirable as it minimizes the need for lab entry; however, these passthroughs can be potential points of inadvertent chemical emission if not properly designed.

- *Maintainability.* Methods are needed for assuring that critical mechanical systems are functional and for facilitating routine maintenance. For example, hood airflow alarms and filter pressure drop sensors are among the design features that should be considered.

- *Decontamination.* Within each working space, it will be important to provide sufficient hood space so that all work can be conducted within hoods, further confining potential contamination. All surface finish (floors, walls, ceiling) must be made of materials that are readily cleaned and that resist reaction or adsorption of chemicals. For this reason, a widely used finish in laboratories is epoxy.

- *Fire protection.* The laboratory should be designed with an integral fire protection system. This requires consideration for the use of sprinklers (and specifications for system design), detectors, alarms and signaling mechanisms.

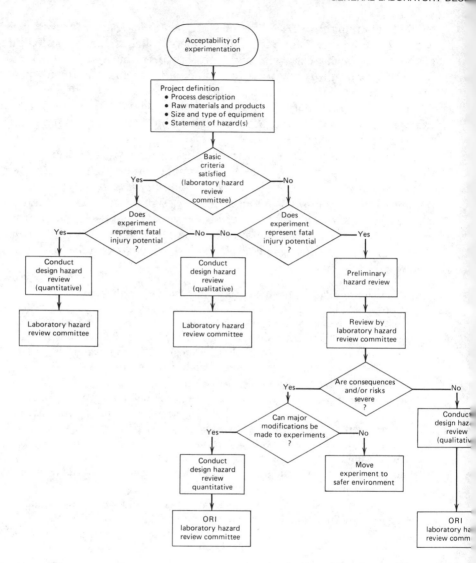

(SOURCE: N. B. Le et al., "Laboratory Safety Design Criteria," presented a
American Institute of Chemical Engineers annual meeting, Boston, July 1986.)

FIG. 8−2: Example of design hazard review process.

TABLE 8−3   Design Considerations

1.0 Laboratory layout
   1.1 Personnel entry and egress
   1.2 Laboratory furniture locations
      Benches
      Aisles
      Desks
      Work surfaces
   1.3 Location of fume hoods
   1.4 Location of equipment
   1.5 Handicapped access
2.0 Laboratory heating, ventilation, and air conditioning (HVAC)
   2.1 Temperature control
   2.2 Laboratory pressure relationship
   2.3 Laboratory ventilation systems
      Comfort ventilation supply air for laboratory modules
      Recirculation of laboratory room air
   2.4 Exhaust ventilation for laboratory modules
      Exhaust of general room ventilation air from laboratories
      Air rates for laboratory hoods and other local exhaust air facilities
      Chemical fume hoods
   2.5 Exhaust fans and blowers
      Exhaust air cleaning for laboratory effluent air
      Exhaust ducts and plenums
3.0 Loss prevention and occupational safety and health protection
   3.1 Emergency considerations
      Emergency fuel gas shutoff
      Groud fault circuit interuppters
      Master electrical disconnect switch
      Emergency blowers
      Emergency eyewash
      Chemical spill control
      Emergency cabinet
   3.2 Construction methods and materials
   3.3 Control systems
   3.4 Alarm systems for experimental equipment
   3.5 Hazardous chemical disposal
   3.6 Chemical storage and handling
   3.7 Compressed gas cylinder racks
   3.8 Safety for equipment

Source: DiBerardinis et al., *Guidelines for Laboratory Design*, Wiley, 1987.

- *Emergency response.* The laboratory should be designed to facilitate emergency response. Means of egress and entrance should be considered, as well as visual accessibility to response teams.
- *Storage.* Provisions must be made for chemical storage within the laboratory. The need of some materials for storage at subambient temperatures must be considered, as should the need for limited exposure to light. The storage area must be secure from exposure to fire and should provide protection against other types of foreseeable accidents (e.g., explosion in an adjacent area).
- *Economic feasibility.* In designing a laboratory, it is possible to build redundancy upon redundancy and minimize risk at great expense. While risk control is of the utmost importance, one must also consider the costs of alternative design features and consider cost effectiveness in design.
- *Ergonomic considerations.* The principles of human factors engineering (ergonomics) must be applied to laboratory design. In a relatively small laboratory, the efficient use of space is important. It is also important to provide a comfortable work environment that minimizes stress, thereby helping to reduce the likelihood of accidents.

Examples of some criteria used to design a special, high hazard containment laboratory are given next. (These criteria would not apply to a general-purpose laboratory.)

Air Supply and Exhaust   The laboratory should have a dedicated air supply system which can provide temperature and humidity control throughout the year. This is important both to provide reasonable working conditions to workers who routinely wear "nonbreathing" protective clothing (e.g., Tyvek) and to protect the experiments from becoming saturated with humidity.

A laboratory's exhaust system should provide a minimum of 10 air changes per hour in each laboratory and storage space. The system should also maintain pressure differentials relative to ambient as follows:

- laboratory hood spaces: −15.2 mm of water,
- chemical and waste storage spaces: −1.52 mm,
- interior laboratory spaces: −0.76 mm, and
- entryways and corridors: +0.76 mm.

Operation of the exhaust system should be monitored and controlled by an automatic system which can maintain these pressure differentials. If a low differential occurs, an alarm should sound both locally and at a central station.

Each laboratory working space should have its own independent exhaust system, including ductwork, fan, filters, and stack. All the hoods, glove boxes, biological safety cabinets, instrument exhausts, and/or animal cage exhausts in a laboratory can be connected to the system for that laboratory. Each of these systems should be connected to a backup that will automatically switch in upon loss, or significant impairment, of exhaust flow in the primary system. Switchover to a backup system should result in the sounding of an alarm, both locally and at a central station.

Lighting: Ergonomic Considerations Lighting should be provided by fluorescent lamps that meet the following standards:

laboratory working and storage areas: 500 lux (50 fc),

bench-top work surfaces: 100−1000 fc,

work surface under hoods: 100−1000 fc,

interior corridors: 300 lux,

lobby and visitor corridor: 100−200 lux, and

changing facility: 500 lux (50 fc).

Fire Protection The entire laboratory complex should be equipped with automatic sprinklers with on/off heads. (It is assumed that the available water supply provides adequate pressure and that a booster pump or auxiliary water supply is not needed.) Each space within the laboratory should have an ionization and a rate-of-rise heat detector for fire detection. Fire alarm pull boxes should be located in the laboratory corridor. The sprinkler flow alarm, pull boxes, and fire detectors should be connected both to an alarm device in the laboratory and to the local fire department. A standpipe and a fire hose reel should be located in the laboratory corridor. Sprinkler flow valves and shutoff valves should be located outside the laboratory.

These design criteria can be utilized to develop several alternative layouts, which can then be rated according to base cost and how well the design meets the design specifications of Table 8−3.

III CHEMICAL HYGIENE PLAN CONSIDERATIONS

The chemical hygiene plan need specify no special design or layout standards for general laboratories. However, it should contain specifications for laboratory equipment, including ventilation controls, as well as personal protective and administrative controls for general laboratories.

REFERENCES

1. L. J. DiBerardinis, et al., *Guidelines for Laboratory Design*. New York: Wiley, 1987.
2. E. R. Hoyle, and R. S. Stricoff, "Functional Requirements and Design Criteria for the Design of High Hazard Containment Laboratories," *Plant/Oper. Prog.*, 6: 3 (July 1987).

RESOURCES

Carter, P., "ANAHL: A Background of Controversy," *Aust. Refrig. Air. Cond. Heat.*, 37:11, 27, 31–32 (November 1983).

Thomas, H. C., "Borrowed Light Reaches Inner Spaces in Lab of the Year," *Res. Dev. (New York)*, 26, 92–99 (May 5, 1984).

Bernheim, F. L. et al., (Architecture Ltd., Chicago), "Complete Design of an R&D Facility — It's a Long Road," *Ind. Res. Dev.*, 24, 112–114 (May 5, 1982).

Gibbons, S.L., and D. C. Davies, "Design of Laboratories for Handling Toxic Substances," *Cent. Toxicol. Lab.* (ICI Ltd.), pp. 467–468.

Gray, W. J. H., "Safety in the Design of Laboratories," *J. Inst. Water Eng. Sci.*, 35:6, 483–90 (1981).

Mellon, M. G., "Some Trends in Planning Chemical Laboratories, Part V, Miscellaneous Trends in Building Materials," *J. Chem. Educ.*, 55:3, 194–197 (1978).

Fried, J., "Meeting Safety Standards in the Laboratory," *Am. Lab.*, 9:13, 79–81 (December 1977).

Mond, C., et al., "Human Factors in Chemical Containment Laboratory Design," *Am. Ind. Hyg. Assoc. J.*, 48:10, 823–829 (1987).

Walters, D., and Jameson, C.W., *Health and Safety for Toxicity Testing*, Lancaster, Pa.: Technomics, 1984.

CHAPTER 9

BARRIER SYSTEM DESIGN

I INTRODUCTION

Modern laboratories designed to handle highly toxic materials incorporate many features not found in traditional laboratory design. Such laboratories are characterized by several areas, depending on the type of testing conducted. These laboratories typically contain areas for chemical receiving, chemical storage, sample (or dose) preparation, and analytical chemistry. Designing these areas requires consideration of the functions and activities to be performed within the barrier systems. A proper design will assure chemical containment.

Designers of containment laboratories must give special consideration to the fire safety implications of barrier and containment features.

This chapter discusses laboratory design, including fire safety and firefighting considerations, within the context of containment laboratories.

II CONTAINMENT OVERVIEW

For modern laboratories, function dictates the particular architectural design and containment control features. For example, some barrier facilities are designed to prevent the accidental release of hazardous agents (e.g., pathogenic microorganisms, toxic chemicals, or radiation). Other barrier facilities are designed specifically to exclude contaminating agents that could compromise a biological experiment (e.g., microbiological experiments or production breeding of experimental animals). Still other barrier facilities may both confine and exclude. This can be achieved through a complementary combination of architectural design features, carefully defined operating practices, specialized safety equipment, and contamination control systems.

No matter what the purpose of the facility, containment of a hazard or exclusion of contamination is achieved through well-established principles of contamination control. There are three levels of containment:

- Primary containment is the protection of laboratory personnel and the environment from direct exposure to hazardous materials. This is provided by engineering controls, such as chemical fume hoods and biological safety cabinets, with the exhaust filtered or treated to remove the contaminants before discharge.
- Secondary containment is the protection of areas external to the laboratory. This is provided by the physical characteristics of the

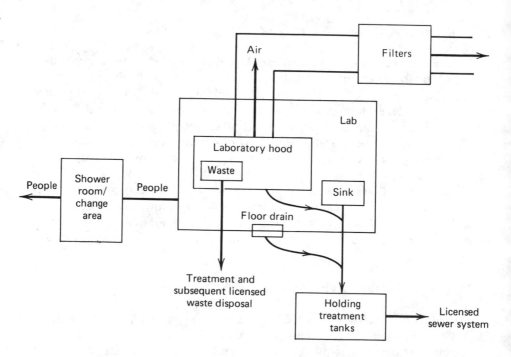

FIG. 9–1: Conceptual schematic for barrier facility design.

laboratory (e.g., corridor and room construction and arrangement, airlocks, ventilation systems, clothing change rooms, showers).

- Tertiary containment exists when an entire area or facility is isolated or physically separated from other structures.

Figure 9–1 provides the overall conceptual design principles in a barrier facility. Decontamination of all personnel, as well as all liquid, solid, and airborne wastes, is done before release from the facility.

It is ironic that containment facilities essential to safeguarding the worker and the environment may become a liability in the event of a fire. As a group, barrier design features tend to contain the heat and toxic products of combustion. They also impede rapid egress and access, hamper search and rescue, delay firefighting and ventilating operations, and prolong overhaul and salvage work. Therefore, barrier design features must also employ fire safety considerations in addition to the chemical containment criteria. Section V addresses the fire safety concerns in barrier system design.

III CHEMICAL HYGIENE PLAN CONSIDERATIONS

A Barrier Systems and Access Restriction

To protect the integrity of experiments and to maximize safe work conditions, the isolation of certain operations is necessary. To meet this objective, any laboratory doing in vivo toxicity testing should adhere to the following criteria:

1. Dose preparation areas are to be isolated from general traffic, either by locating the dose preparation area within the animal facility limited-access barrier system, or by establishing a separate limited-access barrier for dose preparation. If the latter approach is used, all areas into which laboratory workers may bring used protective equipment (including gloves, shoes, head covers, and clothing), respirators, and/or containers of dosed feed or water are to be considered to be behind the barrier. Also, any hallways used by workers for reaching a shower facility are to be considered to be behind the barrier (i.e., a limited-access area).

2. Personnel who enter the dose preparation area, or an area requiring designated protective equipment, must shower out before leaving the barrier facility at the end of the working day.

3. Within the shower facility, the "clean" and "dirty" sides are to be physically separated by the shower or by another physical barrier. The

facility design and procedures are to be arranged so that it is not necessary to cross over to the clean side before showering, or to the dirty side after showering (e.g., to store or retrieve items such as shoes, towels, or respirators) at the end of the work schedule.

4. Each laboratory is to have a room inspection program that provides for monthly checks of the directionality of the airflow; that is:

- relative pressures of laboratory areas are to be checked monthly with smoke tubes to verify that airflows from relatively clean to relatively dirty areas; and
- confirmation of at least 10 and not more than 15 air changes per hour in animal rooms is to be verified at least twice yearly.

B   General

The following design criteria must be used to assure that toxicity testing laboratories provide for proper containment or exclusion of contaminants:

1. If the dose preparation laboratory's general ventilation is partially recirculated, adequate provision for containing the spread of chemicals in the event of a spill should be made. Provisions for spill control will be either the ability to bypass recirculation, exhausting all room air if a spill occurs, or a standard operation procedure (SOP) detailing gas, chemical vapor, and aerosol reduction and chemical clean up procedures.

2. Within the barrier facility, walls, floors, and ceilings are to be sealed around all incoming and outgoing pipes, conduits, and other utilities to prevent release of contaminated material to surrounding areas. Animal and dose preparation rooms are to be constructed of wall, floor, and ceiling materials that form chemical-tight surfaces. Animal room doors are to have windows that permit observation of workers in each room.

3. The relative location of external air intakes and exhausts for both local and general ventilation systems is to be arranged to minimize the risk of reentrainment of exhaust air. Documentation (e.g., a schematic diagram) is to be provided, indicating the location of intakes and exhausts, stack heights, and discharge velocities, as well as the direction of prevailing winds. No weather caps or other obstructions are to be located in the path of vertical discharge.

4. Emergency power generator systems are to be described, including functions and information on maintenance and testing programs (see Chapter 13).

5. Safety showers and eyewash stations are to be located throughout

the facility as required by local, state, and federal regulations, and must be located close to where potentially hazardous materials are used (see Chapter 15, Section III).

IV HOW TO IMPLEMENT CONTAINMENT IN LABORATORY DESIGN

A Receiving and Chemical Storage

The chemical receiving area of a well-designed laboratory is placed away from general traffic, but close to the chemical storage area. The receiving area should be designed to facilitate spill cleanup (e.g., incoming packages that are leaking). An automatic sprinkler system, standpipes, fire extinguishers for hose connections, and fire alarms are provided. A diked area built around chemically resistant floors to contain leaking chemicals is highly desirable. The receiving area should also have adequate storage space. Store chemical containers prone to spill or breakage well away from other activities or equipment.

The bulk chemical storage area should be under negative pressure with respect to adjacent space. Exhaust inlets placed at floor level will prevent the accumulation of heavier-than-air vapors. Electrical fixtures for grounding and bonding flammable liquid drums and cans should also be provided (see Chapter 11).

For storing smaller quantities of flammable liquids, storage cabinets equipped with air-change devices are needed. Freezers and refrigerators for low temperature storage of flammable materials should be suitable for use where fire or explosion hazards may exist. Electrical equipment for such conditions is classified in Article 500 of the National Electrical Code. Manufacturers are required to appropriately label freezers or refrigerators designed for use with flammable materials. The bulk chemical storage room should be equipped with an automatic alarm system. Fire alarms must be easily distinguished from other alarms (e.g., devices indicating temperature or functional anomalies in freezers or incubators).

Proximity of the chemical storage area to the sample preparation area in the barrier facility is desirable. Bulk chemicals should not be transported through the entire barrier facility. Access for bulk chemicals should be either directly to or close to the preparation area.

B Shower/Change Area Facility

The containment facility should feature a shower/change area, the only space in the facility from which one can exit.

The layout of the locker/change area should include traffic patterns that preclude workers from moving back and forth between the clean and dirty sides of the barrier. The shower/change area should have clean and dirty sides, with pass-through showers to allow removal of noncontaminated undergarments (socks, underwear). An area used to store and to clean personal respirators is also needed. Cleaning requires a sink and space for drying. Clean side to dirty side airflow in the locker room is also necessary.

Workers in the barrier facility must wear disposable garments, and the environment should be kept relatively cool and dry to minimize discomfort. A room on the clean side, where workers can spend their break periods without wearing respirators, gloves, and other outer layer protective equipment, is also a worthy design feature.

C Sample (or Dose) Preparation and Working Laboratories

Preparation and working laboratories should be located in the containment facility. Barrier system design must limit the potential spread of chemical contamination. The two-corridor configuration often used in animal laboratories is a recommended design to restrict spead of contamination. Figure 9−2 provides a conceptual overview of a simple barrier facility and Figure 9−3 is an example of a more complex barrier system design.

The sample preparation area of a laboratory, where undiluted test chemicals, unknowns, and positive controls are handled regularly, offers the most potential for personnel exposure to chemicals. The preparation area should be located close to the chemical storage area to minimize contamination throughout the facility. Vented enclosures are needed for weighing and mixing bulk chemicals. The sample preparation area must be under negative pressure and equipped with a static pressure gauge to confirm the existence of the desired pressure differential.

Figure 9−4 diagrams the layout of a sample preparation facility. Bulk chemical storage facilities are located within the dose preparation area to simplify transfer of neat chemical. This preparation area is located in a containment facility with a pass-through/interlock to the clean side of the barrier for transfer. A supplied-air breathing system for handling highly toxic chemicals is included.

Study room doors should have windows for observation purposes. Sealed concrete or a monolithic material, such as heat-welded PVC, is used for floors. The walls are painted with an epoxy. If a computer is needed for data collection and review in a study room, it should be covered with a material that allows decontamination with a suitable solvent. Alternatively, locate the hardware in a clean corridor outside the study room, but allow the workers inside the laboratory access to data input.

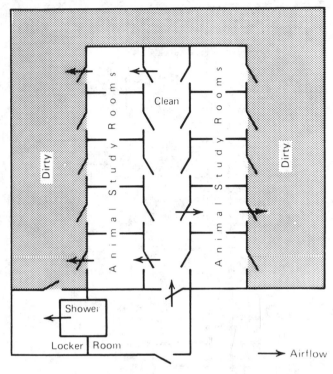

FIG. 9—2: Conceptual example of a simplified two-corridor barrier facility.

D Necropsy and Histology

Where necropsy and histology work are performed, an exhausted work station is needed and test chemical because of the potential for exposure to formaldehyde and other solvents (see Chapter 12, Fig. 12—7, for an example of such a work station). In designing a laboratory, carefully consider the method of supplying formaldehyde to the laboratory. For example, a piped-in system, although it may be operationally convenient, calls for continual vigilance to prevent exposure from leaking fixtures.

Histology work often requires the use of unusually large quantities of flammable solvents. Therefore, provide appropriate flammable liquids storage areas and vented enclosures for equipment used with flammable solvents. Also, because of the fire hazard in an histology area, a fire detection and suppression system should be considered. A fire detection and suppression system (e.g., an automatic carbon dioxide deluge system) would be particularly favorable within a vented enclosure containing

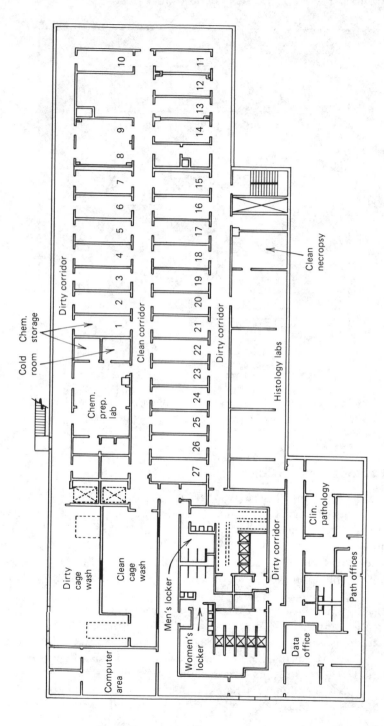

FIG. 9–3: Example of an operating barrier facility (Southern Research Institute).

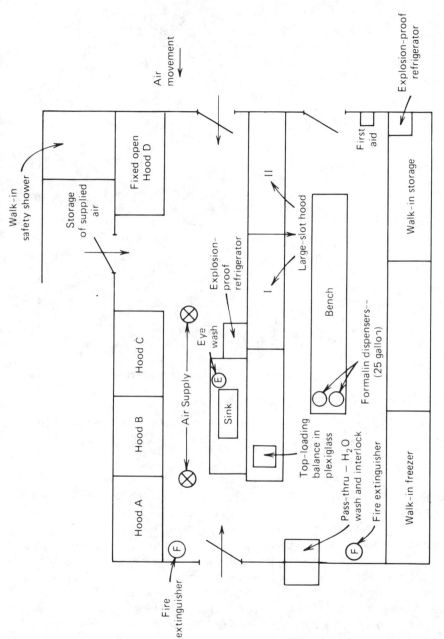

FIG. 9-4: Example of a dose preparation area.

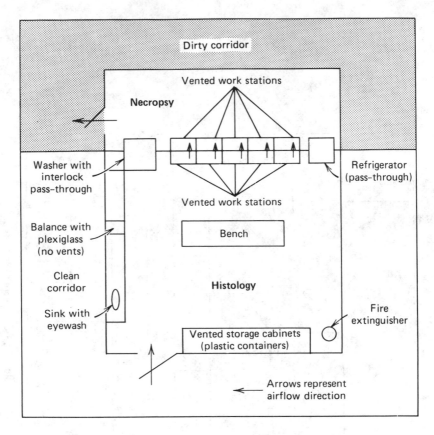

FIG. 9–5: Example of a necropsy–histology laboratory.

automatic staining and processing equipment (see Chapter 12, Section IV for explanation).

Configure the necropsy laboratory to accommodate a clean–dirty barrier. Many times daily animal deaths and large-animal necropsies require the necropsy area to be on the dirty side of the barrier. This occurs because all animals cannot be fully decontaminated before necropsy. The ideal necropsy facility allows necropsies on the dirty side, with tissues and organs passed through to a clean side histology laboratory. Figure 9–5 illustrates a necropsy laboratory with pass-through hoods to achieve clean–dirty separation.

E   Chemistry Laboratories

Equip chemistry laboratories with fume hoods, properly ventilated and positioned (consult Chapter 12). A standard laboratory hood is inadequate for certain operations (e.g. extractions that require especially tall apparatus). These operations call for specially configured, properly designed hoods. Although chemical contamination must be contained when weighing, vibrations and air currents in chemical and biological safety cabinets may disrupt accurate weighing. (Figure 12–6 depicts an alternative method of weighing chemicals in a vented enclosure.) Vent analytical equipment that produces a volatile effluent, and provide an adequate amount of chemical storage. In addition, provide eyewash stations and a safety shower in each work area.

F   General Considerations

The following additional considerations may be appropriate to the design of a containment facility, depending on its primary and secondary purposes:

1. Walls and ceilings should be flat and monolithic, with lighting recessed to minimize places where dust can accumulate. All wall, floor, and ceiling penetrations should be carefully sealed, both to contain the spread of fire and to preserve biological separation between adjacent areas.

2. An isolated, vented, secured storage area for wastes that are later to be removed from the site for incineration or other disposal should be provided, with an interim storage area made immediately accessible to the dirty side of the barrier facility.

3. A cage dump station vented to the outside should be provided where animal cages are dumped before washing.

4. Appropriate utilities support should be provided (e.g., vacuum breakers and accessible traps; drains for collecting contaminated water connected to holding tanks to allow the option of water treatment). Eyewashes and safety showers should be located throughout the facility.

5. Utility shut offs should be located outside the barrier area for emergency response.

6. A breathing air source using an oilless compressor with low pressure alarm, and filters at the compressor and each connection point, should be provided. Alarm indicators are needed in each room where personnel use the compressed air lines, when the compressor is remotely located.

7. A control panel with gauges to indicate ventilation status and with

shut offs for these systems should be provided. This panel should be located outside the barrier and should be readily accessible to maintenance personnel.

8. Electrical fixtures that are waterproofed to allow cleaning by hose should be available. This feature requires floor drains to contain contaminant.

9. The facility should have appropriate fire protection based on anticipated hazards, and emergency means of egress should be provided in accordance with the Fire Safety Code of the National Fire Protection Association (NFPA).

10. An incinerator capable of burning both chemical and animal wastes should be provided relatively close to the laboratory. It should feature an adjustable operating temperature and should comply with all local and state permit requirements.

11. Exhaust air filtered with both high efficiency particulate air (HEPA) and charcoal filters should be provided. Exhaust fans should be located outside the facility so that the positive pressure ductwork that follows the fan is not indoors. Taps should be provided in the filter housing to allow monitoring of air before and after the charcoal filter, to determine breakthrough.

12. Exhaust stacks should be elevated above the roof level to preclude reentrainment, and air intakes should be located on the upwind side of the building rather than on the roof.

13. An emergency electrical generator should be connected to the air-handling systems and some lighting fixtures.

14. A properly alarmed air-handling system to permit rapid detection of fan failures, filter cloggings, and similar malfunctions should be provided.

V FIRE SAFETY AND FIREFIGHTING CONSIDERATIONS IN CONTAINMENT LABORATORY DESIGN

A General Fire Safety Considerations

Both designers and occupants of barrier facilities should consider the possibility of fires and explosions. With regard to such risks, designers can address the following areas:

- *Architectural design features*. Such features as pipe chases and plenums, open stairwells, decorative facades, long, complex corridors, and windowless construction make fire extension more probable.

They also make access and ventilation difficult, and fire suppression more dangerous.

- *Interior finish materials.* Such items as wall and floor coverings (carpeting, drapes, paneling), coatings (paints and varnishes), wallboard, cabinetry, doors, ceiling tile, and insulating materials, although not normally the first items ignited, can and do contribute extensively to the spread of a fire.

- *Construction defects.* Barrier facilities are complex structures with multiple subsystems. Because they are built by workers with various levels of skills, under pressure to complete the job on time, and because of the uncertain quality of inspection, barrier facilities are always prone to construction defects. Unsealed penetration in floors, walls, and ceilings, and missing fire stops or dampers, are examples of defects that make a barrier facility more vulnerable to fire.

- *Fire detection and protection.* The means to detect and suppress fires in laboratories are discussed in Chapter 11.

B Firefighting Considerations

When designing toxicity testing laboratories, consider the strategies that firefighters use in an emergency. Important factors are access, rescue, location, ventilation, salvage, and overhaul. These are addressed in the following subsections.

1 Access

An effective firefighting operation depends on ready access to the premises, not only to the interior, but to the structure itself. Parked service and employee vehicles, building setbacks, landscaping, and high voltage lines can all hinder the approach and deployment of emergency fire vehicles and equipment. If the equipment cannot be brought close enough to the fire, aerial ladders and platforms cannot be raised. Even ground ladders and hose lines may have to be carried inordinately long distances to the conflagration.

Building security measures can restrict access to the interior of a building. Interior access can also be hindred by windowless construction, decorative facades, solar panels, and signs that cover large areas of a building's exterior. With obstacles of this type, firefighters find it difficult to mount their attack. The fire can become well established, possibly spreading into the extensive network of concealed spaces found in a barrier facility.

2 Rescue

Life safety and rescue are the prime commitments of a firefighting force arriving at a building fire. Other important firefighting tasks, such as suppressing the fire and ventilating the building, are relegated to second-alarm personnel who arrive later. Under such circumstances, the action of an automatic sprinkler system is invaluable.

3 Locating the Fire

Locating a fire in a complex barrier facility can be difficult because of the maze of rooms and corridors and the ventilation airflow patterns. Also, in a windowless building, the number of floors may not be readily apparent. If a fire occurs during normal business hours, occupants will be on hand to direct the firefighting force. However, at off-hours locating a fire becomes more difficult. One way to avoid such a problem is to provide the local fire department with building plans showing the locations and types of hazardous and combustible materials, utility shutoffs, service valves, fire protection systems, control valves and panels, and emergency power outlet. This information, along with well-marked stairwells to indicate floor or wing location, gives the firefighters an advantage in mounting an attack.

4 Ventilation

Ventilation controls removal of smoke, gases, and heat from a building. It serves the following functions:

- Protects life by removing or diverting heat, smoke, and toxic gases from locations where building occupants may have found temporary refuge.
- Improves the environment in proximity to the fire by removal of smoke and heat. Removal of heat and increased visibility enable firefighters to advance closer to the fire and extinguish it with a minimum amount of water and time, thus minimizing damage.
- Controls the spread or direction of a fire by establishing air currents that cause the fire to move in a desired direction.
- Allows the release of flammable combustion products before they can accumulate to levels that could create backdraft (explosion) conditions.

Ventilation design features that facilitate firefighting in a barrier facility include the installation of:

- smoke vents, panels, or skylights that open automatically during the initial stages of a fire;
- pressurized stairwells that keep routes of egress free of smoke; and
- doors that close automatically to confine smoke and fire.

5 Salvage

Salvage, the protection of property not already involved in a fire, should commence with the arrival of the fire department. Property can be protected from smoke or water by tarpaulins, or better still by removal from the building. After the fire has been extinguished, water and debris are removed from the property. To facilitate salvage operations, building designs can provide drains or scuppers to channel water used for extinguishment away from expensive laboratory equipment. When hazardous materials are involved, salvage may include removal of such materials, or decontamination operations.

6 Overhaul

The overhaul activity is a thorough investigation of the damaged structure after the main body of the fire has been extinguished to locate — and then extinguish — small pockets of fire still burning in concealed places. The operation involves extinguishing glowing brands and embers that might later rekindle, and opening up all partitions, floors, ceilings, and roofs to look for fire extension. Barrier laboratories with interstitial construction need extensive overhaul operations unless they are well isolated and protected (e.g., with automatic sprinklers).

VI MAINTAINING STRUCTURAL INTEGRITY

To maintain the stability of a structure and the integrity of its interior configuration, the designer must stipulate adequate fire-resistant materials for partitions, ceilings, and support members with cognizance of their fire-loading characteristics (i.e., the type, quantity, and distribution of items that will burn). This design effort will succeed *only* if the designer is knowledgeable of the changes that occur in the physical and chemical characteristics of structural materials when subjected to elevated temperatures. For example, prestressed concrete loses 20% of its strength at 600°F and is permanently weakened at 800°F. Masonry loses its integrity at about 800°F because of joint calcination (loss of water). These characteristics will certainly influence a building designer when considering the fire protection aspects of the project.

TABLE 9–1   Design Deficiencies Responsible for Spread of Fire, Heat, And Smoke

Throughout a Building

Lack of adequate vertical and/or horizontal fire separations.

Unprotected or inadequately protected floor and wall openings for stairs, doors, elevators, escalators, dumbwaiters, ducts, conveyors, chutes, pipe chases, and windows.

Concealed spaces in walls and above ceilings without adequate fire-stopping or fire divisions.

Combustible interior finish, including combustible protective coatings and insulation.

Combustible structural members (beams, girders, and joints) framed into fire walls.

Improper anchorage of structural members in masonry bearing walls.

Explosion or pressure damage to the building because of a lack of or inadequate explosion venting where required.

Damage to unprotected framing resulting in weakening or destruction of floors and walls used as fire barriers.

Lack of means to ventilate fire gases.

From One Building to Another

Lack of adequate fire division walls between adjoining buildings.

Unprotected or inadequately protected openings in fire division walls between adjoining buildings or in fire walls between detached buildings.

Exterior walls with inadequate fire resistance.

Inadequate separation distance.

Combustible roofs, roof coverings, roof structures, overhanging eaves, trim, etc.

Lack of protection at openings to passageways, pipe tunnels, conveyors, ducts, etc., between detached buildings.

Explosion or pressure damage to adjoining or detached buildings.

Collapse of exterior walls.

Source: Walters, D. B., and Jameson, C. W., Health and Safety for Toxicity Testing, Technomics Publishers, Lancaster, Pa., 1984.

Design and construction deficiencies are the main contributors to the spread of a fire, heat, and smoke through a building. Table 9–1 summarizes the deficiencies that can affect structural stability.

VII   SITE SELECTION

In addition to designing a containment facility that offers good access in time of emergency, other site selection factors should be considered. These include space requirements, potential hazards to the community, how climate and terrain can affect hazards, type and size of the facility, special buildings needed, ease of transportation, and availability of a qualified labor supply.

Another important consideration is the proximity of an adequate water supply, both to suppress fire and to control and cover exposures to external fires. Exposures are especially important because laboratory buildings may be located close to other experimental stations or facilities where chemicals, compressed gas cylinders, fuel gas tanks, and hazardous waste are stored. Thus, when planning a barrier facility, provide for the spacing needed to protect each of the adjacent facilities in the event that one or more of them become involved in a fire.

Site planning must also consider potential impacts on the environment, should the facility suffer a fire or explosion. For instance, a laboratory's potential for impacting rivers, streams, sewers, bathing areas, and public parks must be considered. Also, consider the potential for contaminating various waterways in the event of the release of hazardous chemicals and materials during or after a fire occurs. For example, a major fire engenders considerable water runoff. If this runoff becomes contaminated with a hazardous material and the runoff flows into nearby sewers and/or streams, the potential for a major public disaster exists. To preclude such an event, in its flammable liquids code, the NFPA has mandated that containment dikes or tanks be provided for some installations, and that they be large enough to contain both the liquids stored and the potential runoff from a firefighting activity as well. This requirements serves two purposes: it protects the environment, and it prevents exposures to fire involvement.

RESOURCES

General Topics

Accident Prevention Manual for Industrial Operations, 7th ed. Chicago: National Safety Council, 1974, p. 368.

America burning: A Report of the National Commission on Fire Prevention and Control. Washington, DC: Government Printing Office, 1973.

Brannigan, F. L., *Building Construction for the Fire Service*. Quincy, MA: National Fire Protection Association, 1971, Chapter 2.

Chatigny, M. A., and D. I. Clinger, "Contamination Control in Aerobiology," in *An Introduction to Experimental Aerobiology*, L. Dimmick and A. B. Akers, Eds. New York: Wiley, 1969, pp. 194–263.

Clark, W. E., *Firefighting Principles and Practice*. New York: Donnell., 1974, Chapter 5.

Fire Protection Handbook, 15th ed., G. P. McKinnon, Ed. Quincy, MA: National Fire Protection Association, 1981, Sections 5–7.

Fire Protection Through Modern Building Codes. Washington, DC: American Iron and Steel Institute, 1971.

"Flammable Liquids Code," NFPA Standard 30, National Fire Codes. Quincy, MA: National Fire Protection Association, 1984.

Handbook of Industrial Loss Prevention, 2nd ed., Factory Mutual Engineering Corporation. New York: McGraw-Hill, 1967, Chapter 3.

Kuehne, R. W., "Biology Containment Facility for Studying Infectious Diseases," *Appl. Microbiol.*, 26, 239–245 (1973).

"Life Safety Code," NFPA Standard 101, National Fire Codes. Boston: National Fire Protection Association, 1980.

Managing Fire Services, J. L. Bryan and R. C. Picard, Eds. Washington, DC: International City Management Association, 1979, pp. 168–194.

Phillips, G. B., and R. S. Runkle, "Design of Facilities for Microbial Safety," in *CRC Handbook of Laboratory Safety*, 2nd ed., N. V. Steere, Ed. Boca Katon, FL: CRC Press, 1971, pp. 618–632.

Runkel, R. S., and G. B. Phillips, *Microbial Containment Control Facilities*, New York: Van Nostrand Reinhold, 1969, p. 25.

U. S. Department of Health, Education and Welfare. "Design Criteria for Viral Oncology Research Facilities," DHEW Publication (NIH) 75–891. Washington, DC: Government Printing Office, 1975.

Laboratory Design

Further information on health and safety in laboratory design may be found in the following publications:

Code of Federal Regulations, Title 29, Section 1910.134, "Respiratory Protection" (Occupational Safety and Health Administration).

"Fire Protection for Laboratories Using Chemicals," NFPA 45–1986. Quincy, MA: National Fire Protection Association, 1982.

Industrial Ventilation, A Manual of Recommended Practice, 19th ed. Lansing, MI: American Conference of Governmental Industrial Hygienists, 1986.

"Life Safety Code," NFPA 101–1981. Quincy, MA: National Fire Protection Association, 1985.

"National Electrical Code," NFPA 70–1987. Quincy, MA: National Fire Protection Association, 1981.

National Fire Codes, Sections 71, 72A, 72B, 72C, 72D, 72E. Quincy, MA: National Fire Protection Association.

CHAPTER 10

CONSTRUCTION MATERIALS

I INTRODUCTION

Many factors influence the design, construction, and equipping of a laboratory: site location and size, laboratory type, task requirements, hazard containment needs, decontamination needs, and cost are important. Since decisions that pertain to building design, construction, and site preparation have far-reaching effects, consultations with an architect, a structural engineer, and an industrial hygienist are imperative. Despite the variety of approaches to constructing and retrofitting laboratory buildings, minimum requirements and common trends should guide the choice of construction materials used laboratories.

II CONSTRUCTION MATERIAL SELECTION

Laboratories should feature the use of non-permeable surfaces for chemical containment. There are several different approaches to the use of construction materials for chemical containment, depending on whether one is engaged in a new design or retrofitting an existing space. This discussion addresses general laboratory health and safety/construction issues as they relate to chemical containment, utilities, lighting, and furniture.

A Constructing Laboratories for Change

Any building that accommodates a laboratory operation should be designed with potential modification in mind. Although a building shell is constructed for life, the operations it contains are likely to change. Thus, a building's interior design and construction materials should allow for operational changes. Therefore, the structure (building shell) should be separated from the services. A recent trend has been to use interior construction materials that can be easily moved. Often, interstitial spaces are used to permit easy modification of utilities as changes in floor plan and laboratory functions occur. Laboratory building spaces should be designed for flexibility in activities and adaptability to physical rearrangement. The use of aluminum studs and dry wall, coated with epoxy paint, rather than wood studs or concrete block, is an example of an adaptable construction plan.

B General Construction Features

1 General

Certain elements of laboratory design and construction are common to most laboratories. Laboratory floors should be impermeable. Seam-welded PVC flooring or epoxy-coated flooring often is used to achieve impermeability. Flooring should be covered and sealed at the edges with plastic finishes. Walls should be washable and preferably coated with an epoxy paint. Bench surfaces must be smooth and impermeable. Washable suspended ceilings are often used to enclose overhead interstitial spaces. Lighting should be flush with the ceiling. Interior walls should have windows for visual contact from corridors with laboratory spaces. The use of unnecessary horizontal construction materials (finishing) should be avoided. Laboratory doors should be equipped with closers and locks and should have a fire rating of 1 hour. Doors and corridors should be oversized to allow access to equipment.

2 Lighting

Lighting should feature fluorescent fixtures and provide the following illumination levels:

- laboratory and storage work areas: 50 foot-candles,(fc),
- bench-top work surface: 100−1000 fc,
- hood work surface: 100−1000 fc,
- corridors: 30 fc,

- lobby: 10 fc,
- changing facility: 50 fc, and
- animal rooms: 50 fc.

3 Utility Access

As mentioned previously, it is well to locate utilities in an interstitial space. This space should permit easy access from uncontaminated areas. Personnel should not have to enter potentially contaminated areas for routine maintenance. During an emergency this positioning of utility cutoffs allows for remote access and reduces potential exposure to personnel. This provides maximum laboratory flexibility and easy access to make repairs or general maintenance. The space can accommodate ducting for heating, ventilation, and air conditioning, and electrical, plumbing, and fixed gas lines.

4 Furniture

Furniture, such as cabinets and benches, should be fire resistant and should have nonporous surfaces (e.g., stainless steel) for easy decontamination and chemical resistance. For maximum flexibility, furniture should not be permanently mounted to concrete walls.

5 Plumbing

Laboratories should be equipped with floor drains (except in animal rooms) and sinks with glass traps. Drain lines from the floor and sink should be made of chemically resistant plastic, and feed lines should be plastic or copper, with plastic preferred. Waste streams should optimally lead to a water cleaning system capable of removing metals, organics, and particulates. Alternatively, wastes can be sent to double-walled fiberglass or corrosion-protected, steel storage tanks. If wastes are sent to the domestic sewage system, precautions must be taken to assure that contaminant levels do not exceed levels permissible in the waste stream.

6 Electrical

Each laboratory should have an adequate number of 110-V and 220-V grounded electrical outlets that are flush-mounted. Rooms containing large quantities of flammable liquids or gases should have explosion-proof fixtures. An emergency backup power generator should be available to prevent outages (refer to Chapter 13, Emergency Power, for further information). Ground leakage protection should also be provided.

7 *The Outer Shell*

The outer shell of the facility should be made of cinder block, concrete, slab, or brick. Roofs should be supported by insulated I-beams or roof trusses. Concrete footings should be used to support walls, or the facility should be built on a concrete slab or foundation.

III CHEMICAL HYGIENE PLAN CONSIDERATIONS

The chemical hygiene plan should describe the features of the laboratory design which contribute to chemical containment and hazard control.

Within the laboratory facility, for example, walls, floors, and ceilings should be sealed around all incoming and outgoing pipes, conduits, and other utilities to prevent release of contaminated material to surrounding areas. High hazard areas, where carcinogens are handled, should be constructed of wall, floor, and ceiling materials that form chemical-tight surfaces. Doors should have windows to permit observation of the workers in each room.

CHAPTER 11

FIRE AND EXPLOSION PROTECTION

I INTRODUCTION

The potential for fire and/or explosion is always present in laboratories. While the probability and consequences associated with such events vary as a function of laboratory design and operations, steps can be taken to minimize potential losses. This chapter contains information that will assist laboratories in providing appropriate fire and explosion protection.

There are four classes of fires, all of which can occur in the laboratory environment:

Class A fires in ordinary combustible materials, such as wood, cloth, paper, rubber, and some plastics;

Class B fires in flammable liquids, oils, greases, tars, oil-based paints, lacquers, and flammable gases;

Class C fires that are engendered by energized electrical equipment; and

Class D fires in combustible metals, such as magnesium, titanium, zirconium, sodium, lithium, and potassium.

It is important to recognize the differences among these classes, and to understand the types of fire hazard present in an individual laboratory facility. The class(es) of fire hazard present strongly influence the choice of fire detection and suppression equipment.

II PORTABLE FIRE EXTINGUISHERS

A Introduction

Fire extinguishers should be conspicuously located where they will be readily accessible and immediately available in the event of fire. The specific type and size of extinguisher should be selected with consideration for the hazards to be protected and the strength of the personnel who might use the extinguishers.

B Extinguisher Selection: Type

Extinguisher selection should be based on the class of hazards to be protected:

Class A   hazards should be protected with extinguishers of the following types: water, multipurpose dry chemical, bromochlorodifluorome-thane (Halon 1211®), and foam or aqueous film-forming foam (AFFF).

Class B   hazards should be protected with extinguishers of the following types: dry chemical, carbon dioxide, bromochlorodifluoromethane (Halon 1211), bromotrifluoromethane (Halon 1301®), foam, or AFFF.

Class C   hazards should be protected with extinguishers of the following types: dry chemical, carbon dioxide, bromochlorodifluoromethane (Halon 1211), or bromotrifluoromethane (Halon 1301).

Class D   hazards should be protected with extinguishers and extinguishing agents that are approved for use on the specific combustible metal hazard (e.g., G-1® powder for magnesium fires, Lith-X for lithium fires).

C Extinguisher Selection: Size

For the majority of laboratory applications, water and AFFF extinguishers should have a capacity of 2.5 gal. Dry chemical, carbon dioxide, bromochlorodifluoromethane (Halon 1211), bromotrifluoromethane (Halon 1301), and foam extinguishers should hold 20–30 lbs. Site selection should reflect consideration of both the hazards and the strength of the personnel who might use the extinguisher.

D Extinguisher Location and Installation

Locate extinguishers *conspicuously*, where they will be readily accessible in the event of a fire. The travel distance the equipment should be to an extinguisher should be a maximum of 30 ft, and located along normal paths of travel, including exits from an area. Preferably, extinguishers should be located close to any known hazard. The top of the extinguisher is to be installed no more than 5 ft above the floor. The clearance between the bottom of the extinguisher and the floor should be no less than 4 ins. Place operating instructions on the front of the extinguisher.

E Inspection and Maintenance

Conduct inspection of fire extinguishers regularly to ensure that they have been properly placed and are operable. Inspectors should determine that an extinguisher

- is in its designated place,
- is conspicuous,
- is not blocked in any way,
- has not been activated and become partially or completely emptied,
- has not been tampered with,
- has not sustained any obvious physical damage or been subjected to an environment that could interfere with its operations (e.g. corrosive fumes), and
- shows conditions to be satisfactory, if equipped with a pressure gauge and/or tamper indicators.

Inspections should be made at least once a month and documented, with records retained for review.

Maintenance of extinguishers involves a complete and thorough examination, which should include the mechanical parts, the amount and condition of the extinguishing agent, and the condition of the agent's expelling device.

Maintenance techniques vary, and inspections should be performed by qualified personnel. Formal maintenance activities should be conducted at least once each year.

In addition to routine maintenance, hydrostatic testing must be performed on extinguishers subject to internal pressures to protect against failure caused by

- internal corrosion from moisture,
- external corrosion from atmospheric humidity or corrosive vapors,
- damage from rough handling,
- repeated pressurizations,
- manufacturing flaws,
- improper assembly of valves or safety relief discs, and/or
- exposure to abnormal heat (e.g., fire).

Hydrostatic tests should be conducted by qualified personnel using proper equipment. Such tests are often performed by firms that sell and service fire extinguishers. A recommended schedule for hydrostatic testing is:

Type of Extinguisher	Frequency (years)
Water, dry chemical, carbon dioxide foam, aqueous film-forming foam	5
Bromochlorodifluoromethane (Halon 1211)	12
Bromotrifluoromethane (Halon 1301)	12

Use tags and seals to record inspection and maintenance checks. A seal is a good indicator of whether an extinguisher has been used.

Keep a record of the date of purchase and maintenance dates for each extinguisher, giving

- the maintenance date and the name of the person or agency who performed the maintenance;
- the date when last recharged and the name of the person or agency who performed the recharge;
- the hydrostatic retest data and the name of the person or agency who performed the hydrostatic test; and
- a description of dents that remained after the equipment had passed a hydrostatic test.

III DETECTION SYSTEMS

A Types of Detection System

Although heat, smoke, flame, and combustible gas detection systems are not required by laws or regulations in most laboratory facilities, they can be important to protecting the facility and its contents. There are many types of detection system, and the choice of a system must be tailored to the facility and the hazards present.

The following types of detector are often found in laboratory facilities:

Heat detectors. These devices respond to the convected thermal energy of a fire. They are activated when the detecting element reaches a predetermined fixed temperature, or when a specified rate of temperature change occurs. The former are called fixed-temperature detectors; the latter are rate-of-rise detectors. Some detectors combine both features.

Smoke detectors, both ionization and photoelectric types. Ionization detectors typically respond faster to flaming fires, which produce smaller smoke particles. The larger smoke particles generated by smoldering fires are typically detected faster by photoelectric detectors.

Flame detectors. These detectors respond to radiant energy from flames, coals, or embers. There are two types: infrared and ultraviolet flame. The major difference is the insensitivity of the latter to sunlight.

Combustible gas detectors. These units detect the presence of flammable vapors and gases. They give a warning when concentrations in the air approach the explosive range.

B Detection System Selection and Installation

Base the selection of a detector on the anticipated hazard(s) and the environment to be protected. Criteria include type and amount of combustibles, possible ignition sources, environmental conditions, and property values.

Heat detectors are used most effectively to protect confined spaces, or the areas immediately contiguous to a particular hazard. This is because the heat from a fire can dissipate quite rapidly over a larger area, allowing further propagation of fire before the device is tripped. The operating temperature of a heat detector is typically 25°F above the maximum ambient conditions.

Smoke detectors typically respond to fire more quickly than heat detectors and can be used effectively in large, open spaces. Photoelectric devices are preferable if smoldering fires are anticipated, while ionization devices are more effective at detecting flaming fires. Prevailing air currents, as well as ceiling and room configurations, are a key consideration in placement.

Flame detectors are typically installed in high hazard areas where rapid fire detection is critical. Since, infrared flame detectors can experience interference from solar radiation, however, concern for false trips is important. Furthermore, since flame detectors are line-of-sight devices, an unobstructed view of the flame must occur for detection.

Combustible gas detectors are selected and calibrated for the specific substances to be detected. They are typically located close to the hazard and are set to activate an alarm when a certain percentage of the lower flammable limit is reached.

Additional details on automatic fire detector selection and installation can be found in NFPA Standard 72E, "Automatic Fire Detectors."

C Inspection and Maintenance

Inspection and maintenance of detection systems and their components are keys to reliable operation. These activities also help reduce the number of false alarms. It is optimal if such actions are performed regularly and are documented for review.

It is also desirable to test detection systems periodically to assure that they are in proper working order.

IV SPRINKLER SYSTEMS

A Types of Sprinkler Systems

Sprinkler systems automatically provide water or other fire extinguishing agent to extinguish fires. There are water systems, dry chemical systems, Halon systems, carbon dioxide systems, and foam systems.

Water Systems

- *Wet pipe systems*. These are characterized by the presence at all times of water in the lines under pressure. The water will flow through any sprinkler head(s) that fuse (i.e., the closure melts) in a fire environment. Sprinkler heads are available with a variety of operating temperatures and should be installed and replaced only by qualified personnel.

- *Dry pipe systems*. These are characterized by the presence of air in the sprinkler lines under pressure. The air will flow through any head(s) that fuse in a fire environment. This allows water to flow into the lines. Water then flows through the fused head(s).

- *Preaction systems*. These have air in the lines and have a fire detection system. The detection system operates a valve that allows water to flow into the lines. The system then operates like a traditional sprinkler system when a head fuses.

- *Deluge systems*. These feature continuously open sprinkler head(s). There is also a fire detection system which operates a valve, allowing water to flow into the lines and out the open head(s) when a fire is detected.

Dry Chemical Systems

Dry chemicals are powders that are effective in extinguishing Class A, B, and/or C fires. Advantages include quick knockdown capability and absence of electrical conductivity. Disadvantages include slight corrosivity and difficulty in cleanup.

A fixed dry chemical system consists of the agent, an expellant gas, a means to activate the system (e.g., a flame detector), and fixed piping and nozzles. Designs include both total flooding and local application types.

Halon Systems

Halon 1301 is a halogenated hydrocarbon, bromotrifluoromethane, which is effective in extinguishing Class B and C fires. Extinguishment is accomplished by a chemical reaction. An obvious advantage is that Halon leaves no residue after application. Disadvantages include possible toxic effects when agent concentrations exceed 7%, and from the products of decomposition.

A Halon 1301 system consists of the agent, a release mechanism, a means to activate the system (e.g., heat detector), and fixed piping and nozzles. Designs include both total flooding and local application types. In both cases the installation must be in an enclosed area, or the agent must be a liquefied gas.

Carbon Dioxide Systems

Carbon dioxide is effective in extinguishing Class B and C fires. Extinguishment is accomplished by reducing the oxygen content of the atmosphere until it will no longer support combustion. This gas can also extinguish a fire by cooling. Advantages of carbon dioxide include its own pressure for discharge and the lack of residue after use. Disadvantages include need for retention of the extinguishing atmosphere and inherent danger (through oxygen displacement) when used in areas occupied by personnel.

A carbon dioxide system consists of the agent, a means to activate the system (e.g., heat detector), and fixed piping and nozzles. Designs include both total flooding and local application types. An enclosure is mandatory in the former and preferable in the latter.

Foam Systems

Several different types of foam are used to suppress fires and/or vapors from spills of flammable or combustible chemicals. Foams are defined by their expansion ratio, or their final foam volume compared to their original foam solution volume before adding air. There are low expansion ($< 20-1$), medium expansion ($20-200-1$), and high expansion ($200-1000-1$) foams. Foaming agents include aqueous film-forming foam, fluoroprotein foam, alcohol-type foam, and high expansion foam. Application can be effected via fixed systems or through the use of special portable extinguishers.

B   Inspection and Maintenance of Sprinkler Systems

Inspection and maintenance are critical to the reliability of sprinkler system operation. Items to be inspected include the sprinkler control valves, the water pressure or extinguishing agent pressure, and (in the case of dry pipe systems) the air pressure. Fire pumps and suction tanks should also be checked if they are system components. Sprinkler system maintenance should address head condition, corrosion, and freezing. In water systems, periodic flushing of yard mains and branch lines will help ensure reliable water flow. More specific maintenance items are a function of system design (e.g., annual trip testing for dry pipe valves).

V   HANDLING AND STORAGE OF FLAMMABLE AND COMBUSTIBLE LIQUIDS

A   Introduction

Most laboratory operations involve the use of flammable and/or combustible liquids. General safe handling procedures for these materials are summarized below.

All nonworking quantities of flammable liquids should be stored in a storage cabinet approved by Underwriters Laboratories or Factory Mutual, or in a designated flammable liquids storage room with suitable fire protection, ventilation, spill-containment trays, and electrical equipment meeting the requirements of OSHA.

In either storage arrangement, the flammable liquids should be segregated from other hazardous materials such as acids or bases.

Whenever flammable liquids are stored or handled, ignition sources must be eliminated. This means that smoking must be prohibited.

If flammable liquids must be kept at low temperatures, they should be stored in explosion-proof refrigerators.

Flammable liquids transfer should be done in the designated storage room or over a tray within an effective fume hood. In a storage room, all transfer drums should be grounded and bonded and should be equipped with pressure-relief devices and dead-man valves.

Safety cans should be used when handling small quantities of flammable liquids.

B   Classification of Flammable and Combustible Liquids

The classification system for flammable and combustible liquids given here is widely accepted throughout the world. It was prepared by the National Fire Protection Association Technical Committee on Classification and Properties of Flammable Liquids, and is found in NFPA Standard 321, "Basic Classification of Flammable and Combustible Liquids."

Flammable Liquids. Flammable liquids have flash points below 100°F (38°C) and vapor pressures not exceeding 40 psia at 100°F (275 kPa at 38°C). They are classified as Class I liquids and may be subdivided as follows:

- Class IA liquids have flash points below 73°F (23°C) and boiling points below 100°F (38°C).
- Class IB liquids have flash points below 73°F (23°C) and boiling points at or above 100°F (38°C).
- Class IC liquids have flash points at or above 73°F (23°C) and below 100°F (38°C).

Combustible Liquids. Liquids with flash points at or above 100°F (38°C) are referred to as combustible liquids and may be subdivided as follows:

- Class II liquids have flash points at or above 100°F (38°C) and below 140°F (60°C).

- Class IIIA liquids have flash points at or above 140°F (60°C) and below 200°F (93°C).
- Class IIIB liquids have flash points at or above 200°F (93°C).

C Flammable and Combustible Liquids: Handling and Storage

The appropriate measures for the handling and/or storage of flammable and combustible liquids are specific to both site and chemical(s). However, there is an ever-present need to minimize uncontrolled vapors and spills. In addition, ignition sources in the area of use should be controlled. Equipment and procedures for control includes the employment of explosion-proof electrical equipment, flame arresters, grounding and bonding, and the prohibition of smoking.

Additional details on flammable and combustible liquids handling and storage can be found in NFPA Standard 30, "Flammable and Combustible Liquids Code."

VI CHEMICAL HYGIENE PLAN CONSIDERATIONS

The chemical hygiene plan should include coverage of fire prevention and fire response.

The laboratory's practices and procedures for flammable and combustible material storage should be described. Approved storage areas should be identified, and requirements for use of special equipment (safety cans, grounding and bonding, etc.) should be explained.

The fire response plan should be included. The nature of this plan will vary, depending on the fire response strategy selected by the laboratory. There may be a formal fire brigade (in which case OSHA's stringent fire brigade requirements must be met). In other cases, the facility plan may be to evacuate and rely on outside professional assistance. Whatever the plan, it should be described so that all employees may become familiar with their responsibilities in the event of a fire.

VII EQUIPMENT LISTINGS AND APPROVALS

Underwriters' Laboratories (UL) and Factory Mutual Research Corporation (FM) are two organizations that test and rate detectors. They affix a label or stamp to units built and tested according to their standards. They also conduct follow-up examinations of the product and the manufacturer's facilities to ensure that quality standards are maintained.

The companies listed here manufacture equipment that has been approved by the Factory Mutual Research Corporation. In every case, the equipment has been subjected to examinations and inspections and been found to satisfy the criteria for approval. The exact models approved are listed in the latest edition of the *Factory Mutual Approval Guide.*

Heat-Actuated Fire Detectors

AFA Protective Systems, New York, New York

Alarm Industry Products, Farmington, Connecticut

Alison Control, Inc., Fairfield, New Jersey

American District Telegraph Co., New York, New York

Ansul Fire Protection, Wormald US, Inc., Marinette, Wisconsin

BICC General Cables, Ltd., Warrington, England

Chemetron Fire Systems, Inc., University Park, Illinois

Edwards Company, Inc., Farmington, Connecticut

Faraday, Inc., Tecumseh, Michigan

Fenwall, Inc., Ashland, Massachusetts

Fire Control Instruments, Inc., Newton, Massachusetts

Fire-Lite Alarms, Inc., New Haven, Connecticut

The Gamewell Corp., Medway, Massachusetts

Honeywell, Inc., Arlington Heights, Illinois

Kidde Automated Systems, Inc., Westlake, Ohio

King-Fisher Co., Wheeling, Illinois

Mine Safety Appliances Co., Pittsburgh, Pennsylvania

Notifer Co., Lincoln, Nebraska

Protectowire Co., Hanover, Massachusetts

Pyrotronics, Cedar Knolls, New Jersey

Simplex Time Recorder Co., Gardner, Massachusetts

Thermotech, Inc., Ogden, Utah

Smoke-Actuated Fire Detectors

Alarm Device Manufacturing Co., Syosset, New York

American District Telegraph Co., New York, New York

Arrowhead Enterprises, Inc., New Milford, Connecticut

Autocall, Shelby, Ohio

BRK Electronics, Aurora, Illinois

Chubb Fire Security, Ltd., Middlesex, England

Dictograph Security Systems, Florham Park, New Jersey
Edwards Company, Inc., Farmington, Connecticut
Electro Signal Laboratory, Inc., Rockland, Massachusetts
Environment One Corporation, Schenectady, New York
Ericsson Sistemi di Sicurezza SpA, Rome, Italy
Fenwall, Inc., Ashland, Massachusetts
Fire Control Instruments, Inc., Newton, Massachusetts
Fire-Lite Alarms, Inc., New Haven, Connecticut
Firetek Corporation, Hawthorne, New Jersey
First Inertia Switch Ltd., Fleet Hants, England
The Gamewell Corp., Medway, Massachusetts
Hochiki America Corporation, Huntington Beach, California
Honeywell, Inc., Arlington Heights, Illinois
Nittan Corporation, Des Plaines, Illinois
Notifer Co., Lincoln, Nebraska
Protectowire Co., Hanover, Massachusetts
Pyrotector, Inc., Hingham, Massachusetts
Pyrotronics, Cedar Knolls, New Jersey
Simplex Time Recorder Co., Gardner, Massachusetts
Wormald International, Ltd., North Ryde, Australia
Wormald US, Inc., Dallas, Texas

Flame-Actuated Fire Detectors

Alison Control, Inc., Fairfield, New Jersey
American District Telegraph Co., New York, New York
Armtec Industries, Inc., Manchester, New Hampshire
Cerberus, Ltd., Mannedorf, Switzerland
Detector Electronics Corporation, Minneapolis, Minnesota
Edwards Company, Inc., Farmington, Connecticut
Fenwall, Inc., Ashland, Massachusetts
Firetek Corporation, Hawthorne, New Jersey
The Gamewell Corp., Medway, Massachusetts
Pyrotector, Inc., Hingham, Massachusetts
Pyrotronics, Cedar Knolls, New Jersey
Safety Systems, Concord, California

Combustible Gas Detectors

Bacharach Instrument Company, Pittsburgh, Pennsylvania
Control Instruments Corporation, Fairfield, New Jersey
Delphian Corporation, Sunnyvale, California
Detector Electronics Corporation, Minneapolis, Minnesota
Gas Tech, Inc., Mountain View, California
Mine Safety Appliances Co., Pittsburgh, Pennsylvania
Rexnord Gas Detection Products, Sunnyvale, California
Sieger, Ltd., Dorset, England

Dry Chemical Systems

Ansul Fire Protection, Wormald US, Inc., Marinette, Wisconsin
Chemetron Fire Systems, Inc., University Park, Illinois
Walter Kidde Division, Kidde Inc., Wake Forest, North Carolina
PyroChem, Inc., Boonton, New Jersey

Automatic Sprinklers, Standard

Angus Fire Armour Ltd., Thame Oxon, England
Astra Sprinklers Ltd., Sherbrooke, Quebec, Canada
Atlas Fire Engineering Ltd., Swansea, Wales
Automatic Sprinkler Corporation of America, Cleveland, Ohio
Central Sprinkler Corporation, Landsdale, Pennsylvania
Firematic Sprinkler Devices, Inc., Shrewsbury, Massachusetts
GW Sprinkler, Glamsbjerg, Denmark
Globe Fire Equipment Company, Standish, Michigan
Gottschalk Feuerschutz, West Germany
Grinnell Fire Protection Systems Company, Inc., Providence, Rhode
    Island
Grinnell Fire Protection Systems Company, Ltd., Rexdale, Ontario,
    Canada
Guardian Automatic Sprinkler Company, Inc., Nashville, Tennessee
Mather & Platt Ltd., Manchester, England
Miyamoto Kogyosho Ltd., Tokyo, Japan
Preussag Aktiengesellschaft Feuerschutz Minimax, Bad Oldesloe, West
    Germany

Quality Sprinkler Devices, Inc., Cherry Valley, Massachusetts
Reliable Automatic Sprinkler Company, Inc., Mount Vernon, New York
Saval-Kronenburg, Breda, The Netherlands
Star Sprinkler Corporation, Milwaukee, Wisconsin
Total Walther Feuerschutz, Köln-Dellbruck, West Germany
The Viking Corporation, Hastings, Michigan
Wormald International Pty Ltd., Waterloo, Australia

Halon Systems

AFA-Minerva, Ltd., Middlesex, England
Ansul Fire Protection, Wormald US Inc., Marinette, Wisconsin
"Automatic" Sprinkler Corporation of America, Cleveland, Ohio
Cerberus-Guinard Zone, Buc, France
Chemetron Fire Systems, Inc., University Park, Illinois
Chubb Fire Security, Ltd., Toronto, Ontario, Canada
Cronin Fire Equipment, Ltd., Missisauga, Ontario, Canada
FIREBOY Systems, Grand Rapids, Michigan
Fenwal, Inc., Ashland, Massachusetts
Fike Metal Products Corporation, Blue Springs, Missouri
The Walter Kidde Company, Ltd., Middlesex, England
Walter Kidde Division, Kidde Inc., Wake Forest, North Carolina
PyroChem, Inc., Boonton, New Jersey
Pyrotronics, Cedar Knolls, New Jersey

Carbon Dioxide Systems

Ansul Fire Protection, Wormald US, Inc., Mariette, Wisconsin
Chemetron Fire Systems, Inc., University Park, Illinois
Chubb Fire Security, Ltd., Toronto, Ontario, Canada
The Walter Kidde Company, Ltd., Middlesex, England
Walter Kidde Division, Kidde Inc., Wake Forest, North Carolina
Reliable Fire Equipment Company, Alsip, Illinois

Foam Systems

Alison Control, Inc., Fairfield, New Jersey

Feecon Corporation, Westboro, Massachusetts
Foamex Protection, Val d'Or, Quebec, Canada
Walter Kidde Division, Kidde, Inc., Wake Forest, North Carolina
Mine Safety Appliances Company, Pittsburgh, Pennsylvania
National Foam System, Inc., Lionville, Pennsylvania
Rockwood Systems Corporation, Lancaster, Texas
The 3M Company, St. Paul, Minnesota

Water-Filled Extinguishers (2.5 gal.)

Amerex Corporation, Trussville, Alabama
Ansul Fire Protection, Wormald US, Inc., Marinette, Wisconsin
General Fire Extinguisher Corporation, Northbrook, Illinois
Walter Kidde Division, Kidde, Inc., Mebane, North Carolina
Potter-Roemer, Inc., Cerritos, California

Dry Chemical Extinguishers (2.5 lb minimum)

Amerex Corporation, Trussville, Alabama
Ansul Fire Protection, Wormald US, Inc., Marinette, Wisconsin
Walter Kidde Division, Kidde, Inc., Mebane, North Carolina
Pem-All Fire Extinguisher Corporation, Cranford, New Jersey
Potter-Roemer, Inc., Cerritos, California

Multipurpose Dry Chemical Extinguishers (2.5 lb minimum)

Amerex Corporation, Trussville, Alabama
Amway Corporation, Ada, Michigan
Ansul Fire Protection, Wormald US, Inc., Marinette, Wisconsin
Walter Kidde Division, Kidde, Inc., Mebane, North Carolina
Potter-Roemer, Inc., Cerritos, California
Sears Roebuck & Co., Chicago, Illinois

Carbon Dioxide Extinguishers (5 lb minimum)

Amerex Corporation, Trussville, Alabama
General Fire Extinguisher Corporation, Northbrook, Illinois
Walter Kidde Division, Kidde, Inc., Mebane, North Carolina
Potter-Roemer, Inc., Cerritos, California

Vaporizing-Liquid (Halon) Extinguishers (1.0 lb minimum)

ASP International, Inc., Cleveland, Tennessee
Amerex Corporation, Trussville, Alabama
Ansul Fire Protection, Wormald US, Inc., Marinette, Wisconsin
Walter Kidde Division, Kidde, Inc., Mebane, North Carolina
Metalcraft, Inc., Baltimore, Maryland
Potter-Roemer, Inc., Cerritos, California

Air Foam Extinguishers (2.5 gal minimum)

Amerex Corporation, Trussville, Alabama
Ansul Fire Protection, Wormald US, Inc., Marinette, Wisconsin

Dry Compound Extinguishers

Ansul Fire Protection, Wormald US, Inc., Marinette, Wisconsin

Steel Storage Cabinets

A&A Sheet Products, Inc., LaPorte, Indiana
CAH Industries, Inc., Elk Grove Village, Illinois
Eagle Manufacturing Company, Wellburg, West Virginia
Equipto, Inc., Aurora, Illinois
Justrite Manufacturing Company, Mattoon, Illinois
Kewaunee Scientific Equipment Corporation, Statesville, North
   Carolina
Labconco Corporation, Kansas City, Missouri
Protectoseal Company, Bensenville, Illinois
Trojan Metal Products, Inc., Los Angeles, California
The Williams Bros. Corporation, Scarborough, Ontario, Canada

Bonding and Grounding Assemblies

Stewart R. Browne Manufacturing Company, Inc., Atlanta, Georgia
Centryco, Inc., Burlington, New Jersey
Protectoseal Company, Bensenville, Illinois

Safety Bungs (Pressure-Relief Devices)

CAH Industries, Inc., Elk Grove Village, Illinois
Central Illinois Manufacturing Company, Bement, Illinois

Centryco, Inc., Burlington, New Jersey
Justrite Manufacturing Company, Mattoon, Illinois
Protectoseal Company, Bensenville, Illinois

Self-Closing Faucets (Dead-Man Valves)

CAH Industries, Inc., Elk Grove Village, Illinois
Central Illinois Manufacturing Company, Bement, Illinois
Centryco, Inc., Burlington, New Jersey
Conbraco Industries, Inc., Matthews, North Carolina
Dyko Manufacturing, Inc., Manchester, New York
Economy Safety Faucet Company, Lansdale, Pennsylvania
Imperial Clevite, Inc., Chicago, Illinois
Justrite Manufacturing Company, Mattoon, Illinois
Protectoseal Company, Bensenville, Illinois
H. B. Sherman Manufacturing Company, Battle Creek, Michigan

Safety Cans

Brookins, Inc., Mishawaka, Indiana
Champ Service Items, Edwardsville, Kansas
Cooper Industries, Inc., Canfield, Ohio
John Deere Merchandise, Inc., Moline, Illinois
Eagle Manufacturing Company, Wellsburg, West Virginia
Hercules Can Company, Mattoon, Illinois
Justrite Manufacturing Company, Mattoon, Illinois
KP Industries, Delphos, Ohio
New Delphos Manufacturing Company, Delphos, Ohio
Protectoseal Company, Bensenville, Illinois
TRW, Inc., Cleveland, Ohio

CHAPTER 12

LOCAL EXHAUST VENTILATION

I INTRODUCTION

Any laboratory working with hazardous materials must install and operate adequate ventilation.

Two common ventilation problems are commonly found in laboratory health and safety reviews. First, many hood monitoring and maintenance programs concentrate only on visible components of the system (e.g., the hood), ignoring such ancillary equipment as ductwork and fans. Second, many scientists work with poorly designed ventilation systems.

II USE OF LOCAL EXHAUST VENTILATION IN LABORATORIES

Three types of local exhaust ventilation (LEV) system are commonly found in laboratories: chemical fume hoods, biological safety cabinets, and miscellaneous exhausted enclosures.

A Chemical Fume Foods

The common chemical fume hood is simply an exhausted enclosure that draws hazardous materials (vapors, fumes, dusts, and mists) away from the breathing zone of the laboratory worker. It is typically equipped with a horizontal or vertical sash so that the hood opening can be minimized during use and closed when not in use. Makeup air can be introduced directly into the enclosure or taken from the ambient air. The laboratory chemical fume hood derives its effectiveness from the fact that to the greatest extent possible, the operation of concern is enclosed, and air is being mechanically exhausted away from personnel.

B   Biological Safety Cabinets (BSC)

There are three classes of biological safety cabinets in common use in laboratories. The following convenient nomenclature, established by the National Sanitation Foundation, is used in the descriptions of the various classes and types. Generally speaking, BSCs of Classes I, II, and III are used for work that involves substances of low toxicity, moderate toxicity, and high toxicity, respectively. These devices are described and illustrated next.

Class I. A Class I BSC is similar to a common laboratory chemical fume hood except that the exhaust is filtered through a high efficiency particulate air (HEPA) filter. There is no recirculation of filtered air back into the work area. The airflow characteristics of a Class I BSC are illustrated in Figure 12−1.

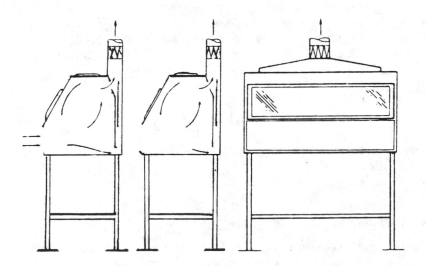

FIG. 12−1:   Class I biological safety cabinet.

Class II. A Class II BSC is designed so that a downward-directed laminar flow of air exists during operation, while an inward flow of air is maintained at the cabinet opening. Four types of Class II BSCs have been defined; the differences among them have to do with the amount of air recirculation and the manner in which exhaust air is removed from the work area.

Type A. These, cabinets recirculate 70% of the HEPA-filtered exhaust air from the work area. The remaining 30% is exhausted through an exhaust duct and may reenter the general laboratory environments. Type A cabinets are inappropriate for use when gases or vapors that have passed through the HEPA filter are present. Since greater than half of the air is recirculated, these gases or vapors will tend to accumulate within the cabinet. A typical airflow pattern for a Class II, Type A BSC is shown in Figure 12–2.

Type B1. Type B1 biological safety cabinets (Fig. 12–3) differ from type A in that only 30% of the exhausted air from the work area is recirculated. Since less than half the air is recirculated, it is a better choice of BSC when toxic or hazardous gases or vapors may be present.

Type B2. None of the air exhausted from the work area is recirculated (100% exhaust) inside the enclosure of a Type B2 BSC. All the air, after having passed through a HEPA filter, is discharged, either into

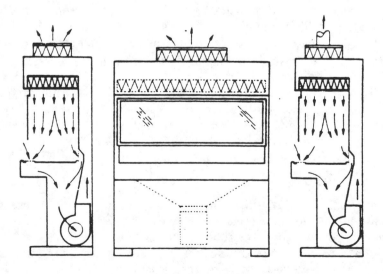

FIG. 12–2: Class II, Type A biological safety cabinet.

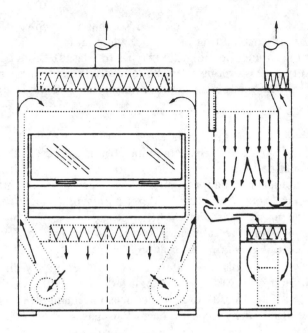

FIG. 12−3:   Class II, Type B1 biological safety cabinet.

an appropriate plenum and duct or back into the workroom air. The vertical laminar flow is induced by a second blower dedicated to supply air. This configuration is illustrated in Figure 12−4. Class II, type B2 cabinets are also referred to as 100% total exhaust cabinets.

Type B3. The airflow on a Class II, type B3, cabinet is identical to that of a type A cabinet (Fig. 12−2) in that 70% of the air exhausted from the work area is recirculated. Like the Type A configuration, it is inappropriate for work involving toxic or hazardous gases or vapors. The major differences between the two types include a higher minimum face velocity on the Type B3 (100 vs. 75 fpm), and the requirement that the 30% non-recirculated air be exhausted outside the building.

Class III. A Class III BSC is a gas-tight enclosure used for operations with the highest level of risk. As shown in Figure 12−5, access to the interior is possible only via rubber sleeves and gloves that have been integrated into the side of the cabinet. Air is swept across the work area through HEPA filters on both the supply and exhaust ducts.

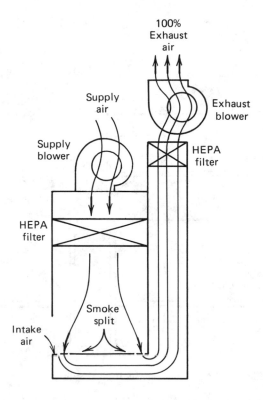

FIG. 12—4:   Class II, Type B1 biological safety cabinet.

C   Other Exhaust Enclosures

Other special systems for laboratory facilities also available include small dosing hoods, recirculation desktop hoods equipped with charcoal filters for use with solvents, walk-in hoods for large equipment, "elephant trunks" for gas chromatographic and spectrophotometric equipment effluents, and extraction hoods used for large chemical extraction apparatus.

Specialized exhausted enclosures can be designed to accommodate both task requirements and contamination control needs. The exhaust enclosure for an analytical balance shown in Figure 12—6 is an example of such coordinated design. This type of exhaust enclosure addresses the concerns of potential chemical exposure in weighing operations. Design criteria that were incorporated into the balance enclosure include easy access, a rear exhaust for effective air distribution, and allowance of enough area for transfer of the neat chemical inside the enclosure.

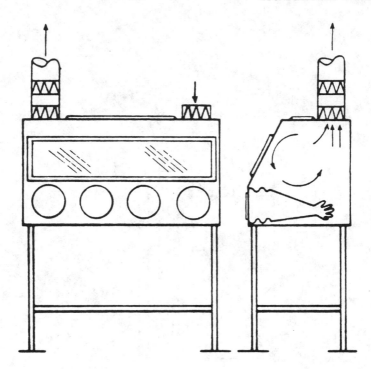

FIG. 12−5:   Class III biological safety cabinet.

Local exhaust ventilation systems, typically, are composed of five basic components: the hood, the ductwork, an air cleaner, a fan, and an exhaust stack (Fig. 12−7). The hood is the most apparent part of the system, but the other four components should be considered just as critical; a breakdown in any one can compromise the whole system.

III   DESIGN OF VENTILATION SYSTEMS

The design of an effective and efficient ventilation system is a technically complex and rigorous process, and all designs should be developed or approved by a registered professional engineer or a certified industrial hygienist. The design process requires detailed familiarity with the range of tasks for which protection is required, as well as with the variety of biological and chemical agents to be used in the system. The following are guidelines that should be considered when designing a system:

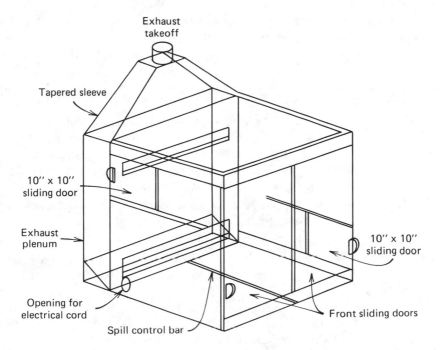

FIG. 12−6:   Prototype vented balance enclosure.

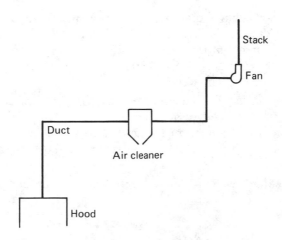

FIG. 12−7:   Local exhaust ventilation system.

1. When designing a local exhaust ventilation (LEV) system for a variety of applications, be sure that it will be protective enough for the least easily controlled operation and the most toxic biological or chemical agent.

2. Locate hoods in a manner that minimizes the amount of traffic that can disrupt the flow patterns of a hood.

3. Enclose as much of the operation or process as possible. Not only does this increase the effectiveness and reliability of control, but it also reduces the amount of air that has to be handled. Equip the enclosure with horizontal or vertical sashes, so that the hood opening can be minimized when in use and shut completely when not in use.

4. Avoid placing the hood near doors, windows, or air diffusers that can create cross-drafts at the hood face.

5. Be sure that any makeup or auxiliary air introduced at the hood face flows in the same direction as the induced draft into the hood.

6. Round the front edge or lip of the hood opening to reduce turbulence.

7. Design the duct connection to the hood to avoid sharp angles of introduction of other branches to within six duct diameters of the hood. This will reduce turbulent air patterns inside the hood.

8. Be sure that the volume of air exhausted from the hood is sufficient to achieve a capture velocity of 100 ±20 ft/min of the hood opening. If the operation in the hood results in a certain volume of contaminated gas or vapor, the exhaust volume should be large enough to accommodate that volume.

9. Use air-cleaning equipment that is appropriate for the type of contaminant. The equipment should be located so that filters can be easily changed or monitored. Locate differential pressure gauges in the laboratory so that pressure drops, indicative of the need to replace filters, can be easily determined.

10. Locate the fan outside the building (on the roof), to permit the maintenance of negative pressure in the duct carrying contaminated air from the hood to the fan. Locate air-cleaning equipment ahead of the fan to reduce deterioration of fan parts caused by the action of the contaminants.

11. Place exhaust stacks above the roof line and away from air supply equipment.

IV HOOD MONITORING AND INSPECTION PROGRAMS

The performance criteria listed below reflect both quantitative and qualitative assessments of ventilation system performance and condition. Any system used to protect personnel or the environment from toxic or hazardous substances should be rigorously tested before being put into service — and frequently thereafter. When evaluating the performance of a given system, the following criteria should be considered:

1. nondisruptive interior and exterior air patterns, including those in the hood or cabinet, as well as those from the room environment into the blood;
2. adequate and appropriate capture velocities and airflow volumes;
3. absence of leaks from plenums, sashes, hoods, or from the gloves in a glove box;
4. adequate routine and preventive maintenance;
5. use of pressure gauges and alarms to determine the pressure drop across the air-cleaning devices; and
6. housekeeping.

All laboratories performing work involving hazardous materials, where exhaust ventilation is used for primary control of personnel exposures, should operate according to a regularly scheduled ventilation system monitoring and maintenance program. Such a program should be designed and directed by qualified health and safety personnel. Because of the wide variety of laboratory hoods (chemical fume hoods, biological safety cabinets, vented enclosures) and other local exhaust ventilation (e.g., vented waste containers, refrigerators), the monitoring program for each piece of exhaust equipment should reflect the manufacturer's recommended operating practices. As described below, each monitoring program should include daily visual inspections, quarterly inspections, annual maintenance and testing, and user training.

A Daily Visual Inspection Before Operation

Exhaust Slots. Adjustable rear exhaust slots in laboratory hoods should be checked periodically for proper balance. When the exhaust inflow and downflow are unbalanced in a biological laminar flow cabinet, the potential exists for contaminated air to be pushed outside the hood face.

Airflow check. When the exhaust system is operating (hoods, any vented equipment), a tissue paper check should be performed to ensure that the exhaust is functional: tape a small piece of tissue paper at the hood opening and observe whether it reflects a directional airflow. For a glove box, tissue paper should be placed inside at the exhaust slots (but it should not be allowed to escape into the exhaust system, as it might block the filter and reduce airflow).

Smoke test. All exhaust enclosures should be smoke-tested to demonstrate the effective capture of contaminants generated during normal operating procedures. The smoke test is conducted by placing a smoke generator inside the enclosure and observing whether all the generated smoke is captured. If smoke leaks out of the enclosure, contaminated air could peak out during normal conditions.

Pressure gauges. All pressure gauges should be checked to see that they indicate pressure levels with a predetermined safe operating range. This range should be initially determined for both the enclosure static pressure and the HEPA filter pressure drop by a qualified safety and health professional. The precise range may vary between systems, and thus it should be defined on a system-by-system basis.

Housekeeping. Any material blocking the hood opening or exhaust parts should be removed. If evidence of turbulence has been found during the check of the hood, one explanation may be clutter in the hood. Hoods are not designed to be chemical storage cabinets.

B   Quarterly Inspections

Smoke tube tests. Regularly scheduled smoke tube tests serve to evaluate irregular or turbulent airflow patterns. Tests should be performed as follows:

1. *Chemical fume foods or Class I biological safety cabinets* (with or without auxiliary air supply). The smoke should move from several inches in front of the sash directly to the rear exhaust slot, with the sash in its normal operating position. The smoke tube should be placed at and above the interior working space to locate any dead or turbulent spots.

2. *Class II biological safety cabinets.* The smoke should move from several inches in front of the sash into the forward intake grille. The

smoke at the working area should move toward either the forward or rear exhaust slots with a minimum of turbulence.

3. *Glove Boxes or Class III biological safety cabinets.* Since these installations are gas-tight and maintained under negative pressure, a smoke tube test is performed to identify leaks around the joints and seals. The test is conducted by placing the smoke tube at the outside glove gaskets and inside the rubber gloves, and checking for evidence of smoke inside the box.

4. *Other exhaust equipment.* The smoke tube should be placed around the perimeter or the outer boundaries of the area to be exhausted (e.g., the edge of a square waste container vented along the opposite edge). The smoke should move directly to the exhaust slots.

Face or Capture Velocity. Hood face velocities and laminar downflows should be measured quarterly with a velometer. These test should be performed by qualified personnel using a properly calibrated thermal or mechanical velometer.

1. *Chemical fume hoods.* The average of the velocities in each of six theoretical sections (Fig 12−8) at the hood face will represent the overall face velocity. On the basis of prior velocity measurements, a specific sash height should be designated and marked as the *safe operating position.* This is the sash position at which a satisfactory face velocity is achieved; the hood must not be operated with an opening greater than that indicated by the safe operation sash position.

Subsequent velocity measurements should also be taken at the safe operating sash position to allow for a common basis for comparison. When these measurements are made, any auxiliary air should be turned off. If the hood is connected to other hoods, the face velocity should be measured under the maximum worst conditions (e.g., the connecting hoods with sashes fully opened and exhaust on). The face velocity should be 100 ± 20 fpm, with no individual point less than 80 or greater than 120 linear feet per minute.

2. *Biological safety cabinets (laminar flow).* In Class II hoods, Types A and B, the supply blower should be switched off for the face velocity measurement. The face velocity should be 100 fpm ±10%. The vertical downflow of the supply blower should also be measured. The downflow at the work area should be approximately 50−80 fpm, depending on the manufacturer's recommendations.

Several types of Class II, Type B, hoods are not designed for easy

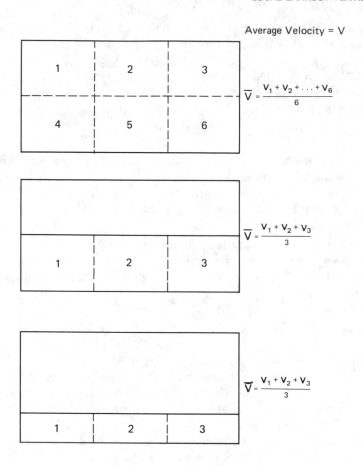

FIG. 12–8: Hood face velocity measurement technique.

switchoff of the supply blower for face velocity measurements. For these hoods, a combination of supply–working surface measurements and inlet smoke tubes tests should be performed by experienced personnel. If the inlet supply air is too great (e.g., > 80 fpm) and the smoke tubes indicate lazy inflow air patterns at the face, the supply–exhaust airflow may be out of balance; alternatively, the HEPA filters could be overloaded, or the exhaust fan could be malfunctioning. Whatever the problem, further testing or maintenance may be required.

3. *Glove box*. If the glove box has a filtered air inlet with no blower, a velometer reading can be taken to determine the exhaust volume. (The velometer can be placed directly at the air inlet to measure the inlet

velocity. This velocity times the inlet area will give the exhaust air volume). This exhaust volume for most manufacturers' specifications is in the range of 30–50 cfm (cubic feet per minute). When a glove box has a supply air blower (consequently, the inlet air volume is not available), a combination of periodic smoke tube/leak tests and annual exhaust velocity duct measurements should be performed as discussed above.

C   User Training

The daily visual inspections and smoke tube tests should be an important aspect of a hood operator's education and training. Before an employee begins using any laboratory hood, he or she should be trained in its proper use and become familiar with monitoring programs. The results of these tests for all types of exhaust equipment should be recorded and the records kept in an accessible location so that hood users can refer to them when suspected hood malfunctions occur. A sample hood monitoring form is shown Figure 12–9.

Room: _____   Date: _____
Location within room: _____
Sash: Yes: _____   None: _____
          Vertical movement: _____   Horizontal movement: _____
Damper: Yes: _____   None: _____
Fan switch: Yes: _____   None: _____
Apparent connection to other hoods:
Sketch (Show position of hoods, benches, windows, doors, makeup air.)

Face velocity measurements (Test with other hoods on.)

Condition/Position	Top Left	Bottom Left	Top Center	Bottom Center	Top Right	Bottom Right	Average
Sash open, door open							
Sash open, door closed							
Sash at operating level, door open							
Sash at operating level, door closed							

Hood checked by: _____

FIG. 12–9:   Sample hood monitoring form.

D Routine Maintenance

In addition to the periodic monitoring, an overall annual maintenance procedure should be established. These procedures should be performed by *qualified* personnel.

Documentation of these tests should be maintained by laboratory supervisory personnel (Figs. 12−9 and 12−10 are examples of hood monitoring forms).

The following general maintenance procedures should be performed annually on exhaust equipment. More specific information for special hoods and enclosures is presented in the National Sanitation Foundation Standard for Class I Biohazard Cabinets, NSF 49.

Laboratory _____   Model _____

Investigator _____   Serial No. _____

Telephone No. _____   Date _____

Supply blower speed
control setting _____   Supply Velocity

Magnehelic gauge
reading _____" w.c.

High ____ Low____ Average Work
Surface_____

Air Curtain

Access Opening _____

Exhaust Velocity   Ducted _____

Not Ducted ____
Size _____

High __ Low__ Average__

CFM ____ ____
Calculated Air Intake _____

REMARKS:_____
_____
_____
_____

Testing Technician _____

FIG. 12−10: Sample velocity profile from biological safety cabinet (laminar flow).

Date _____   Company _____
Unit Identification   *Unit Location*
  Type _____   Building _____
  Model No. _____   Room _____
  Serial No. _____   Other _____

Tests Performed
  ( ) Velocity profile
  ( ) Calculated inflow
  ( ) In-place leak testing of HEPA filters (single-particle monitor)
  ( ) In-place leak testing of HEPA filter (DOP method)
  ( ) Paraformaldehyde decontamination
  ( ) Halogen leak test
  ( ) Ground continuity — containment and intrusion
  ( ) Smoke tube test
  ( ) Noise level & vibration test

Final Testing Results   ( ) Unit failed tests
  ( ) Unit passed tests   ( ) Requires new filters
    ( ) All leaks repaired   ( ) Requires frame repair
    ( ) New filters installed   ( ) Other
    ( ) Other

*REMARKS:*_____
_____
_____
_____

                           Testing Technician _____

FIG. 12–10: Continued

- *Exhaust fan maintenance.* The fan manufacturer should recommend necessary maintenance, including lubrication, belt checking, fan blade deterioration, and speed check recommendations. Lubrication of the fan and fan motor may be required more frequently, depending on the operating conditions.

- *Ductwork check.* All the ductwork between the hood or exhaust opening should be checked for corrosion, deterioration, and buildup of liquid or solid condensate. Any dampers used for balancing the system should be lubricated and checked for proper operation. Unused ductwork or old hood installations should be removed.

- *Air-cleaning equipment.* In-line exhaust charcoal or HEPA filters should be monitored for contaminant buildup. Mechanical or absorbent filters not equipped with differential pressure gages, or audible alarms, should be leak-checked annually. Absorbent or absorbent filters for gas and vapors can be leak-checked by the release of halogen gas inside the hood and a halogen meter monitoring the filter outlet airstream. HEPA filters can be checked using the dioctyl phthalate (DOP) method (see NSF Standard 49); this type of test is often performed by an outside contractor.

- *Velocity measurement.* As mentioned earlier, total exhaust Class II, Type B biological safety cabinets and glove boxes require exhaust duct velocity measurements to verify the proper airflow in the enclosures. [In the total exhaust hood, this supply volume (cfm) can be substracted from the exhaust volume (cfm), and this value can be divided by the area of the face opening to calculate the face velocity.] For the glove box, the exhaust duct volume should fall within the range of 30−50 cfm.

When these maintenance procedures are done, extra precautions should be taken to protect personnel from any of the toxic contaminants in the hood, ductwork, or filters. Any excess contaminated material or filters removed from the hood system should be disposed of according to the facility's approved toxic waste disposal practices.

E Special Tests for Biological Safety Cabinets

Figure 12−10 mentions several special tests that are used to maintain and evaluate biological safety cabinets. Some of these tests may be performed by the manufacturer; however, the purchasing laboratory should test each cabinet to ensure containment, since shipping, relocation, filter, load, and installation of the cabinet may alter performance. Testing should conform to the National Sanitation Foundation requirements outlined in NSF Standard 49. These NSF testing procedures for velocity profile, plenum leak testing (halogen/soap bubble), HEPA filter leak testing, and personnel, product and cross-contamination protection check (biological challenge) are included as Appendix A to this chapter.

V FIRE-RELATED ISSUES

An additional fire-related issue involves the exhausting or ventilation procedure associated with solvent-containing apparatus left in operation overnight. Over the past several years, fires have occurred in some of these units as a result of sparks igniting the vapor. In at least one instance during a power failure, a tissue processor in a histopathology laboratory was turned off in the open position, which allowed fumes to escape into the enclosure. When the power returned, sparks from the immersion heater in the paraffin bath ignited the vapor.

As a result of these incidents, several steps are recommended before unattended operation of solvent-containing equipment is initiated. If emergency power backup is available, the risk of fire is greatly reduced. However, if emergency power is not available, a night watchman or security person should be alerted to turn the equipment off before start-up. For information on other fire-related issues, the reader is referred to Chapter 11.

VI   CHEMICAL HYGIENE PLAN CONSIDERATIONS

The chemical hygiene plan should describe the facility's local exhaust ventilation program. The plan should describe the types of operation for which local exhaust ventilation is to be used, the proper use of the ventilation, the inspection, testing, and maintenance program for local exhaust systems, and the documentation of the program that will be maintained.

RESOURCES

The following sources can provide more detailed information on ventilation monitoring programs for specific types of laboratory hoods and exhausted enclosures:

Caplan, K. J. and G. W. Knutson, "The Effect of Room Air Challenge on the Efficiency of Laboratory Fume Hoods," *ASHRAE Trans.*, 83, Part I, 141–156 (1977).

Chamberlin, R. I., and J. E. Leahy "A Study of Laboratory Fume Hoods," U.S. Environmental Protection Agency Report, Contract 68–01–4661, 1978.

Fuller, F. H. and A. W. Etchells, "The Rating of Laboratory Hood Performance," *ASHRAE J.*, pp. 49–53, October 1979.

Gaffney, L. F. et al., "Field Testing and Performance Certification of Laboratory Fume Hoods," presented at Industrial Hygiene Conference, May 1980.

Industrial Ventilation – A Manual of Recommended Practice, 17th ed., Committee on Industrial Ventilation, American Conference of Governmental Industrial Hygienists.

Cincinnati, OH: American Conference of Governmental Industrial Hygienists, 1973.

National Sanitation Foundation Standard 49 for Class II (Laminar Flow) Biohazard Cabinetry. Ann Arbor, Mi: National Sanitation Foundation, 1976.

Scientific Apparatus Makers Association Standard LF10−1980, "Laboratory Fume Hoods." Washington, DC: Scientific Apparatus Makers Association, 1980.

Stuart, D. G., M. W. First, R. L. Jones, Jr., and J. M. Eagleson, Jr., "Comparison of Chemical Vapor Handling by Three Types of Class II Biological Safety Cabinets," *Particulate Microb. Control*, 2(2): 18−24 (1983).

U.S. Department of Health, Education and Welfare, DHEW Publication 76−162, "Recommended Industrial Ventilation Guidelines," National Institute of Occupational Safety and Health Contract CDC-99−74−33, prepared by Arthur D. Little, Inc., Cambridge, MA, January 1976.

U.S. Department of Labor, Occupational Safety an Health Administration, Subpart 2 − Toxic and Hazardous Substances 1910.1000−1910−1045, "*General Industry*" (rev.). Washington, DC: Government Printing Office, 1978.

APPENDIX A
PERFORMANCE TESTS *

    I. Soap Bubble/Hallogen Leak Test

       A. Purpose (Soup Bubble Test)

        This test is performed on exterior surfaces of all plenums to determine if the welds, gaskets, and workmanship are free of leaks.

       B. Apparatus

        Liquid leak detector, Search or equal, to meet Military Specification MIL-L-25567A.

       C. Procedure

         1. Prepare the test area of the cabinet as per Sections I.G.1 and 2, I.G.2 [below].

         2. Pressurize the test area with air to a reading of 2 inches (50.8 mm) water gage.

         3. Spray or brush the liquid leak detector along all welds, gaskets, penetrations, or seals on the exterior surfaces of the cabinet plenums. Leaks will be indicated by bubbles. Leaks will occur that blow the detection fluid from the hole without forming bubbles. This type of leak must be detected by slight feel of airflow or sound.

       D. Acceptance

        All welds, gaskets, penetrations, or seals on the exterior surfaces of the air plenums shall be free of soap bubbles at 2 inches (50.8 mm) water gauge.

       E. Purpose (Halogen Leak Test)

        This is to be performed on all contaminated air plenums under

---

* From National Sanitation Foundation Standard 49 for Class II (Laminar Flow) Biohazard Cabinetry, National Sanitation Foundation, Ann Arbor, Michigan.

positive pressure to the room. This test is performed to determine if the exterior joints made by welding, gasketing, or sealing with sealants are free of leaks that might release potentially hazardous materials into the atmosphere.

F. Apparatus

The instrument used for detecting halogen leaks shall be capable of detecting a halide lead of 0.5 ounce per year (8.9×10^{-6} cc/sec). The unit shall be calibrated in accordance with the manufacturer's instructions using a calibrated leak.

G. Procedure

1. Prepare the test area of the cabinet as a closed system (i.e., seal the front window opening, exhaust port, removable panels, and/or other penetrations).

2. Attach a manometer, pressure gage, or pressure transducer system to the test area to indicate the interior pressure.

3. Pressurize the test area with air to a reading of 2 inches (50.8 mm) water gage. If the test area holds this pressure without loss for 30 minutes without additional air being supplied, release pressure. (If the test area does not hold this pressure, examine for gross leaks with soap solution or equal.)

4. The room in which the testing will be performed shall be free of halogenated compounds, and air movements shall be kept to a minimum. No smoking should take place in the test area.

5. Pressurize the test area to 2 inches (50.8 mm) water gauge pressure using halide gas (dichlorodifluoromethane).

6. Adjust the sensitivity of the instrument in accordance with the manufacturer's instructions. The nozzle of the detector probe shall be held at the surface of the test area so as not to jar the instrument and should be moved over the surface at the rate of about 1/2 inch (13 mm) per second, keeping probe (6.4−12.7 mm) away from the surface.

7. Move the probe over the seams, joints, utility penetrations, panel gaskets, and other areas of possible leakage. [Fig. 12A−1]

H. Acceptance

Halogen leakage shall not exceed full scale reading at 8.9×10^{-5} cc per second sensitivity (0.5 ounce per year or less at 2 inches [50.8 mm] water gauge pressure).

II. HEPA Filter Leak Test

A. Purpose

This test is performed to determine the integrity of the HEPA filters, the filter housings, and the filter mounting frames. The cabinet shall be operated at the manufacturer's recommended airflow velocities.

B. Apparatus

The instruments shall be:

1. An aerosol photometer with either linear or logarithmic scale.

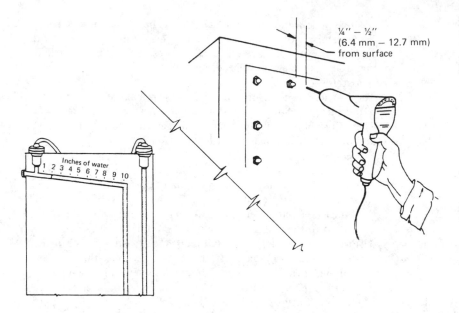

¼″ — ½″
(6.4 mm — 12.7 mm)
from surface

Inches of water
1 2 3 4 5 6 7 8 9 10

FIG. 12A–1: Halogen leak test.

The instrument shall have a threshold sensitivity of at least 1×10^{-3} micrograms per liter for 0.3 micrometer diameter dioctylphthalate (DOP) particles, and a capacity of measuring a concentration of 80 to 120 micrograms per liter. The sampling rate of air shall be at least 1 cfm (28.3 liters/minute).

2. A DOP generator of the Laskin nozzle(s) type. An aerosol of DOP shall be created by flowing air through liquid DOP. The compressed air supplied to the generator should be adjusted to a pressure of 20 psi (although lower pressure can be used) and to a minimum free airflow through each nozzle of 1 cfm (28.3 liters/minute).

C. Procedure

1. Place the generator so the DOP aerosol is introduced into the cabinet upstream of the HEPA filter.

2. Turn on the photometer and calibrate in accordance with the manufacturer's instructions.

3. Measure the DOP concentration upstream of the HEPA filter.

   a. For linear readout photometers (graduated 0–100), adjust the instrument to read 100 percent while using at least one Laskin type nozzle per 500 cfm (14,160 liters/minute) airflow or increments thereof.

        b. For logarithmic readout photometers, adjust (using the instrument calibration curve) the upstream concentration to 1×10^4 above the concentration necessary for one scale division.

    4. Scan the downstream side of the HEPA filters and the perimeter of each filter pack by passing the photometer probe in slightly overlapping strokes over the entire surface of the HEPA filter, with the nozzle of the probe not more than 1 in (25.4 mm) from the surface. Scan the entire periphery of the filter, the junction between the filter, and filter mounting frame. Scanning shall be done at a traverse rate of not more than 2 inches/second (5 cm/second).

  D. Acceptance

    DOP penetration shall not exceed 0.01 percent measured by a linear or logarithmic photometer.

VII. Personnel, Product, and Cross-Contamination Protection (Biological) Tests

  A. Purpose

    These tests are performed to assure that aerosols will be contained within the cabinet, outside contaminants will not enter the work area of the cabinet, and aerosols created within the cabinet will not contaminate other equipment located within the cabinet.* The cabinet shall be operated at the manufacturer's recommended intake and downflow velocities. Cabinets meeting these tests shall then meet airflow characteristics as measured in Sections IX [below] and X [not included in this appendix].

  B. Materials

    1. Spores of *Bacillus subtilis* var. *niger (b. subtilis)*

    2. Sterile diluent prepared as follows:

       Gelatin — 2 grams

       Na_2HPO_4 — 4 grams

       Distilled H_2O — 1000 ml

       Adjust pH to 7.0

       Autoclave at 250 degrees F (121 degrees C) for 20 minutes.

    3. Petri plates (100×15 mm and 150×22 mm) containing nutrient agar, Trypticase Soy Agar (BBL), or other suitable growth medium with no inhibitors or other additives.

    4. AGI-4 samplers (flow rate calibrated at 12.5 liters per minute) containing 20 ml of sterile distilled water with 0.06 percent antifoam. The AGI-4 samplers shall be Ace Glass Incorporated, Vineland, New Jersey, catalog number 7542−10, air sampling impingers or equal.

---

* The success of these biological tests is dependent on the proper relationship between the downflow air velocity, inflow air velocity, and the height of the access opening.

5. Slit type samplers operating at 1 cubic foot per minute (0.000472 m^3/s).

6. Refluxing nebulizer with impingement anvil, Fisons Corporation (2 Preston Court, Bedford, Massachusetts 01730, or equal, operating at a dissemination rate of approximately 0.2 ml/minute at 10 psi (0.70 kg/cm^2).

7. One 2−1/2 inch (63.5 mm) outside diameter stainless steel, steel or aluminum cylinder with closed ends shall be used to disrupt the airflow. The lengths determined by size of interior of cabinet. One end butts against the back wall of the cabinet, and the other end protrudes at least 6 inches (152.4 mm) into the room through the front opening of the cabinet.

C. Personnel Protection Test (System challenged with 1×10^8 to 8×10^8 B. subtilis spores)

1. Procedure

a. A nebulizer containing 5 ml of the spore suspension is centered between the side walls of the cabinet. The horizontal spray axis is placed 14 inches (355.6 mm) above the work surface with the opening of the nebulizer positioned 4 inches (101.6 mm) in back of the front window with the spray axis parallel to the work surface and directed toward the front window. [Fig. 12A−2]

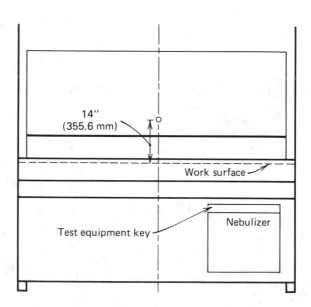

FIG. 12A−2: Personnel protection test.

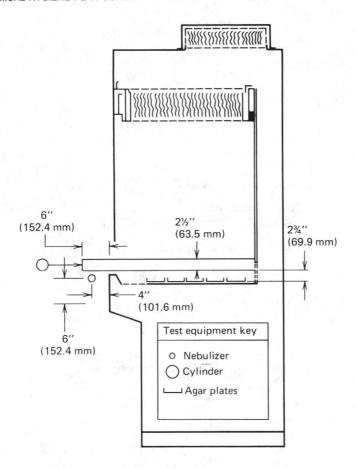

FIG. 12A−3: Personnel protection test.

b. The cylinder is placed at the center of the cabinet. The axis
of the cylinder is 2−3/4 inches (6.9.9 mm) above the work
surface. [Fig. 12A−3]

Around the cylinder, four AGI's are positioned with the
sampling inlets 2−1/2 inches (63.5 mm) outside the front of
the cabinet. Two of the four AGI's are placed so their inlet
axes are 6 inches (152.4 mm) apart and are in a horizontal
plane tangent to the top of the cylinder. Two AGI's are
positioned so their inlet axes are 2 inches (50.8 mm) apart
and lie in a horizontal plane 1 inch (25.4 mm) below the
cylinder. An agar plate is placed under the cylinder at the

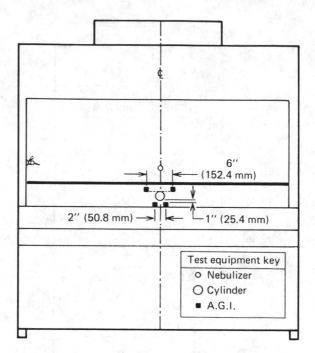

FIG. 12A−4: Personnel protection test.

front edge of the work tray as a positive control plate. [Fig. 12A−4]

c. Two slit type samplers are placed so the horizontal plane of the air inlets is at the work surface elevation and the vertical axes of the inlets are 6 inches (152.4 mm) in front of the cabinet and 8 inches (203.2 mm) from each sidewall. Two slit samplers are placed so the horizontal plane of the air inlets is 14 inches (355.6 mm) above the work surface and the vertical axes are 2 inches (50.8 mm) outside the front edge of the cabinet and 6 inches (152.4 mm) on each side of the cabinet centerline. [Fig. 12A−5]

d. The duration of the test is 30 minutes. The test sequence is as follows:

Time (minutes)	Activity
30	Start slit samplers
16	Start nebulizer
15	Start impingers

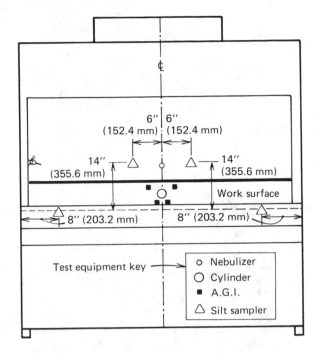

FIG. 12A−5: Personnel protection test.

10	Stop impingers
9.5	Stop nebulizer
0	Stop slit samplers

A total of five replicate tests will be performed.

e. From each impinger, three 0.1 ml samples will be pipetted onto the surface of three 100 × 15 mm agar plates, and spread with a sterile glass or metal spreader. The remaining fluid will be pooled and filtered through 0.22 or 0.45 micrometer membrane filter, the filter aseptically removed, and placed on appropriate media. The plates containing the filters, the 0.1 ml samples, and those from the slit samplers will be incubated at 98.6 deg. F (37 deg. C) for 48 hours. Examine at 24−28 hours and 44−52 hours.

2. Acceptance

The number of *B. subtilis* organisms recovered from the combined collection suspension of the four AGl samplers shall not

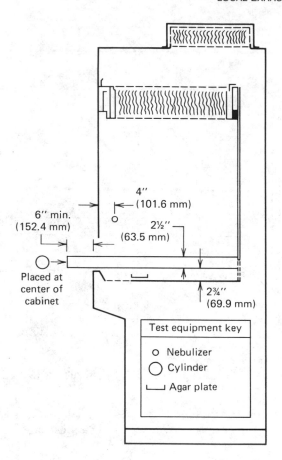

FIG. 12A−6: Product protection test.

exceed 20 colonies per test. Slit sampler plate counts shall not exceed five *B. subtilis* colonies for a 30-minute sampling period. The control plate shall be positive. A plate is "positive" when it contains greater than 300 colonies of *B. subtilis*.

D. Product Protection Test (System challenged by 1×10^6 to 8×10^6 *B. subtilis* colonies in 5 minutes)

  1. Procedure

    a. Completely cover the work surface with open agar settling plates. [Fig. 12A-6]

    b. The horizontal spray axis of the nebulizer is positioned at the level of the top edge of the work opening and is centered between the two sides of the cabinet with the opening of the nebulizer 4 inches (101.6 mm) outside the window. The spray axis is parallel to the work surface and is directed toward the open front of the cabinet.

c. A 2−1/4 inch (63.5 mm) outside diameter cylinder with closed ends is placed in the center of the cabinet. The cylinder is positioned in the cabinet so one end butts against the back wall of the cabinet, the other end extends at least 6 inches (152.4 mm) into the room through the front opening of the cabinet, and the axis of the cylinder is 2−3/4 inches (69.9 mm) above the work surface.

d. Place a positive control plate approximately 6 inches (152.4 mm) below the nebulizer.

e. After nebulization is complete (5 minutes), continue to operate the cabinet for 5 minutes.

f. Place the covers on the open agar plates and incubate at 98.6 deg. F (37 deg. C) for 48 hours. Examine at 24−28 hours and 44−52 hours.

2. Acceptance

The number of *B. subtilis* colonies on the agar settling plates shall not exceed five organisms for each test with at least five replicates. The control plate shall be positive. A plate is "positive" when it contains greater than 300 colonies of *B. subtilis*.

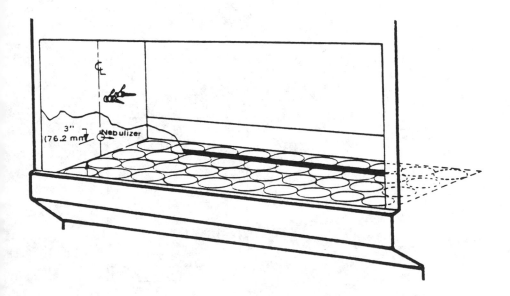

FIG. 12A−7: Cross-contamination test.

E. Cross-Contamination Test (System challenged by 1×10^4 to 8×10^4 *B. subtilis* colonies in 5 minutes)
  1. Procedure
      a. Cover the work surface with rows of open agar settling plates starting against the side wall.
      b. The horizontal spray axis of the nebulizer is positioned 3 inches (76.2 mm) above the work surface and located against the midpoint of the right or left interior side wall, with the opening directed toward the opposite side wall. [Fig. 12A-7]
      c. After nebulization is complete (5 minutes), continue to operate the cabinet for 15 minutes.

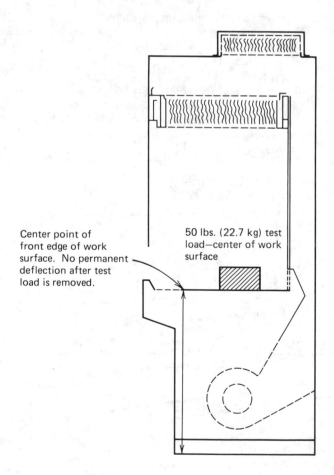

Center point of front edge of work surface. No permanent deflection after test load is removed.

50 lbs. (22.7 kg) test load—center of work surface

FIG. 12A−8: Resistance to deflection.

   d. Place the covers on the open agar plates and incubate at
      98.6 deg. F (37 deg. C) for 48 hours. Examine at 24–28
      hours and 44–52 hours.
2. Acceptance
   Many agar plates will recover spores of *B. subtilis*. These
   plates should form a fan-shaped pattern from the nebulizer,
   but agar plates whose centers are greater than 14 inches
   (355.6 mm) from the side wall shall be free of *B. subtilis*
   colonies. [Fig. 12A-8, Fig. 12A-9]

                            . . .

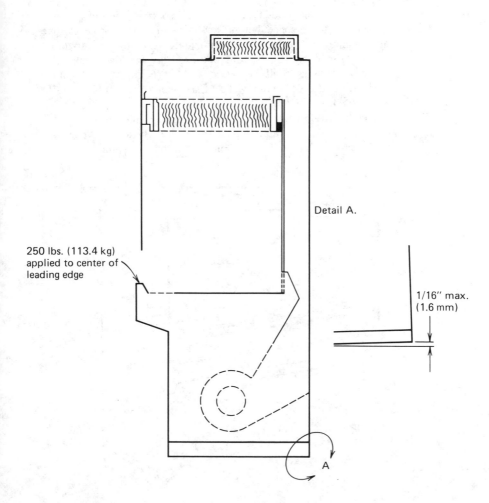

FIG. 12A–9: Resistance to tipping.

IX. Velocity Profile Test
   A. Purpose
      This test is performed to measure the velocity of the air moving through the cabinet work space and is to be performed on all cabinets accepted under the performance test Section VII. Thereafter, all units of such production models shall meet the manufacturer's stated downflow velocities.
   B. Apparatus
      A thermoanemometer with a sensitivity of ± 2 fpm (± .01 m/s) or 3 percent of the indicated velocity shall be used.
   C. Procedure
      Measure the air velocity in the work space at multiple points across the work space below the filters on a grid scale to give approximately nine readings per square foot in the horizontal plane defined by the bottom edge of the window frame. Air velocity readings shall be taken at least 6 inches (152.4 mm) away from the perimeter walls of the work area. [Fig. 12A-10, Fig. 12A-11]
      Particular attention should be given to corners and upper edges of the face access opening. Repeat the scan with the smoke stick 2 to 3 inches (50.8 to 76.2 mm) inside the face opening.

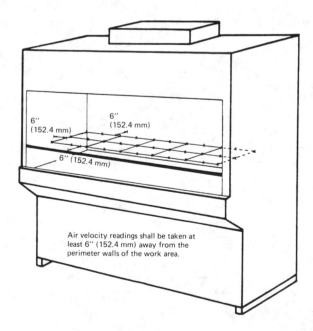

FIG. 12A−10:   Velocity profile test.

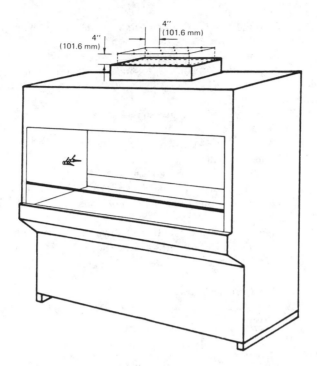

FIG. 12A−11: Work access opening airflow.

D. Acceptance

   Directional airflow as shown by smoke shall be inward through the face access opening into the forward intake grill.

   . . .

XII. Ultraviolet Lighting Intensity Test

   A. Purpose

      This test is performed to determine, in microwatts per square centimeter, the ultraviolet (UV) radiation on the work tray surface in the cabinet.

   B. Apparatus

      1. Portable photoelectric UV intensity meter capable of measuring UV radiation at a wavelength of 253.7 nanometers. It shall be used in accordance with the manufacturer's instructions.

      2. The UV intensity meter will be calibrated in accordance with the manufacturer's instructions.

      3. Provide eye protection for the test personnel.

   C. Procedure

      CAUTION: Turn off electricity before cleaning.

1. Clean the UV tube with a soft wipe moistened with 70 percent alcohol.
2. Turn on UV lights and allow to warm up for 5 minutes.
3. Take readings according to the manufacturer's instructions.
4. Calculate the UV intensity for the work tray surface.

D. Acceptance

UV radiation wave length shall be 253.7 nanometers. The UV irradiation intensity on the work tray surface shall be not less than 40 microwatts per square centimeter.

XIII. Drain Spillage Trough Leakage Test

A. Purpose

This test is performed to demonstrate the containment capability of the spillage trough under the work surface.

B. Procedure

Remove the work surface from the cabinet and fill drain spillage trough with water and hold it for 1 hour. Check for visible signs of water leakage after 1 hour.

C. Acceptance

The drain spillage trough shall hold at least 4 liters of water and have no visible signs of water leakage water 1 hour holding period.

CHAPTER 13

EMERGENCY POWER

I INTRODUCTION

The interruption of utility service to a laboratory facility can cause major adverse effects. Interruptions can be caused by storms, earthquakes, vandalism, maintenance outages, and equipment breakdowns. Each facility should have a comprehensive plan that outlines its response to a loss of utilities. If the loss does not create an emergency, it will render the facility more vulnerable to an emergency situation. Therefore, a utility interruption contingency plan should be included as part of a facility's plan for emergency operations.

II IMPLICATIONS OF POWER OUTAGES

Loss of electricity is more likely than loss of water pressure, and the effects on a laboratory could be more widespread and more serious. In considering the implications of an interruption in electricity, each facility must examine the measures necessary to protect the laboratory, its occupants, and the integrity of any experiments being run.

There should be auxiliary backup power for all emergency equipment. Exit signs, lighting, evacuation alarms, and other emergency equipment must not be disabled at a time when they will be most necessary. Frequent operational checks must be conducted where backup batteries are used to ensure a constant state of readiness. If a system-wide backup generator is used, it must be designed to provide power immediately, without a time delay. In addition, equipment to be used by emergency personnel in the event of a power outage must be available.

Equipment used to control employee exposure to hazardous materials should be kept operating during a power outage, if at all possible. It is desirable to equip air-moving and air-cleaning devices critical to personal safety and health with backup power. In addition, if the effectiveness of the ventilation system could be compromised by a power outage, the rate at which air contaminants are generated must be minimized. To this end, shutdown procedures should be established for key pieces of equipment and experiments.

If auxiliary power is used to maintain the ventilation system, the characteristics of the system running on auxiliary power must be evaluated. Any change in airflow between rooms, or in any operational parameter of the ventilation system, must be recognized. The impact of changes on the safety and health of laboratory personnel must be assessed, and evacuation and emergency response plans adjusted accordingly.

Finally, to the extent possible, it is desirable to provide auxiliary power to ensure the integrity of experiments in progress. Backup power must be provided to all refrigerated areas, and a continuous source of power must be ensured for the computers being used.

Any interruption of utilities may have adverse consequences that place a facility, its employees, and the surrounding community at an elevated risk. In addition, the substantial investment in ongoing experiments could be placed in jeopardy. It is the responsibility of laboratory managers to foresee these consequences and to make adequate preparations to minimize them.

III MAINTENANCE AND TESTING CONSIDERATIONS

A maintenance program is required for all emergency power systems. The maintenance program should include regular testing of emergency equipment (generators, batteries, etc.) as recommended by the manufacturer or by the engineering department. The testing program should be documented and described.

The emergency procedures to be followed in case of power outage should be regularly reviewed and regularly exercised. Regular emergency drills help to assure that laboratory staff will be familiar with equipment shutdown, staff notification, and evacuation procedures on the relatively rare occasions when the procedures are used.

IV **CHEMICAL HYGIENE PLAN CONSIDERATIONS**

The chemical hygiene plan should address loss of power and the appropriate response by laboratory personnel.

- If there is an emergency power generator system, its operations, functions, maintenance, and testing schedule should be addressed.
- If there is no emergency power backup system to supply areas critical to safety and exposure control (such as hoods and exhaust enclosures), emergency shutdown and evacuation procedures to be followed in the event of power loss should be included.

CHAPTER 14

HUMAN FACTORS DESIGN CRITERIA

I INTRODUCTION

In most working environments, the success of a particular endeavor is a function of, among other things, the performance of the persons doing the work. In such cases, it is possible to optimize work effectiveness and probability of success by systematically identifying and eliminating problems in the way that the worker interfaces with the working environment. These problems may include conditions or practices that decrease efficiency or increase the stress and strain imposed on participants. Often, though, a careful analysis of tasks and working conditions will lead to a design, or redesign, of the working environment that will eliminate or minimize the problems and enhance the performance of workers. This process of analysis and design of work spaces is called human factors engineering, or ergonomics.

A second, though not less important, goal of human factors engineering is safety. Frequently, a rigorous analysis of a working environment will identify conditions or practices that create an elevated risk of personal injury or spillage of a toxic or hazardous substance. Poor or inappropriate lighting at dose administration areas, for example, may increase the likelihood of exposure to the dosing agent. Poor design of a chemical handling area may result in preventable spills or accidents. Finally, if the amount of physical force required to accomplish a particular task exceeds the capabilities of the worker, overexertion injuries may result.

II HUMAN FACTORS ENGINEERING TECHNIQUES

A Task Analysis

The analysis of tasks for the purpose of human factors engineering evaluation addresses four characteristics of each task in question:

- range of motion and reach requirements,
- exertion and strength requirements,
- dexterity and fine motor control requirements, and
- frequency and duration.

The range of motion required to perform a task must be compatible with the physical and anthropometric dimensions of the worker. The reach requirements, which are measured in terms of horizontal and vertical distance from the normal working position, can be compared to established data for working populations. The design criteria are created to establish horizontal and/or vertical reach requirements so that 95% of the population from which workers are drawn can comfortably perform the task.

If the task in question requires a reach, or range of motion, that is at or beyond the maximum for the working population, increased effort will be required to perform the task. This may lead to articles being dropped, increased fatigue, or inappropriate task performance.

The amount of exertion required to perform a task is an easily recognizable limitation for workers. However, the amount of weight that can be lifted repeatedly without causing strain or musculoskeletal stress is lower than might be expected. Thus not only is the weight of the object being moved important in assessing the biomechanical stress associated with a task, but the range of motion through which it is moved is also a critical determinant.

Figures 14−1 and 14−2 illustrate the relation between the weight and location of an object to be lifted in terms of the particular compressive force experienced in the back of the person doing the lifting under the weight−location conditions covered. It is clear that the potential damage a lifting task can inflict on a worker's back is a function of both weight and location.

The use of Figures 14−1 and 14−2 is illustrated by the following example. (For the sake of clarity, only the horizontal location of the weight is considered.)

The laboratory safety and health officer wishes to assess the safety of a lifting task in which a 5-gal bucket is filled with water and carried to the animal area. The laboratory spigot from which the bucket is filled is located 25 in. from the side of the sink.

Five gallons of water weighs approximately 41.5 lb (18 kg); this weight is lifted at a horizontal distance of 25 in. (64 cm). Clearly this task will result in a back compressive force greater than the 350 kg recommended by the National Institute of Occupational Safety and Health (NIOSH) for women (see Fig. 14−1), but less than the 650-kg guideline for men (Fig. 14−2).

The second task of carrying the water is also a concern of the safety officer. The diameter of the top of the bucket is 15 in., and the horizontal distance of the load while walking is the radial distance, 7.5 in. (19 cm). By inspection of Figures 14−1 and 14−2, it can be seen that the task of lifting an 8-kg object at a horizontal distance of 19 cm will not generate excessive back compressive force in either women or men.

The safety and health officer, having determined that the task of lifting a 5-gal bucket of water out of a sink may be inappropriate for the persons

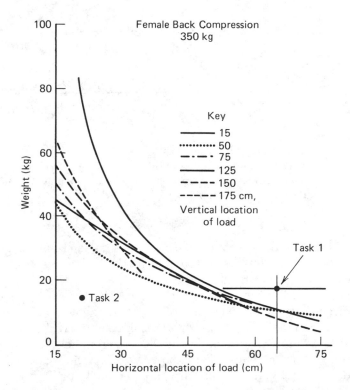

FIG. 14−1: Task variables producing 350 kg — female back compression.
SOURCE: "Work Practices Guide for Manual Lifting," NIOSH Publication 81−112, National Institute for Occupational Health and Safety, Cincinnati, OH 45226, March 1981.

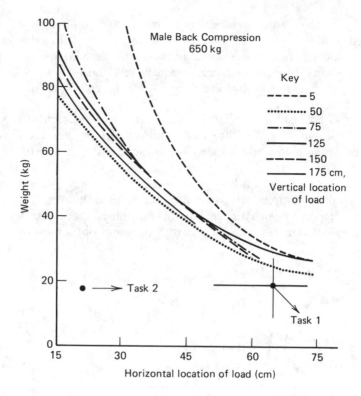

FIG. 14–2: Task variables producing 650 kg — male back compression.
SOURCE: "Work Practices Guide for Manual Lifting," NIOSH Publication 81–122, National Institute for Occupational Safety and Health, Cincinnati, OH 45226, March 1981.)

working in the area, is keeping a cardinal rule of human factors engineering, namely, the first priority is to redesign the job so that it is safe for the person who is to perform it.

Several remedial measures are possible. First, smaller quantities of water can be carried. At a distance of 45 cm, a weight of 11 kg (~ 3 gal of water) can be lifted without excessive back compressive force.

A second solution is to redesign the filling station. If a hose were available, so that the bucket could be filled on the floor and then picked up and carried by the technician, the potential for injury would be commensurately reduced.

The biochemical characterization of a lifting task or other activity involving physical exertion is a complex and rigorous undertaking. It should be attempted only by persons familiar with the process, terminology, and execution of such an analysis. Remedial action, however, is often more straightforward. It can range from using lifting assists to reducing the amount of physical work necessary. The overriding principle is that the task should be appropriate to the person doing it.

Many laboratory tasks require dexterity and fine motor control. The problems associated with these tasks differ from those discussed earlier. Tenseness, eye strain, stiffness, and other complaints often accompany close attention and relative immobility. Here, special consideration must be paid to the work station and other aspects of the work environment.

A final critical aspect of task analysis is consideration of the duration of the task and the frequency with which it must be performed. The detrimental effects from repetitive activities that require significant physical effort, or close attention, can accumulate over time. Thus, the worker should be given opportunities to pause and rest when necessary. Since

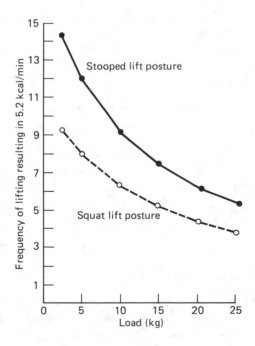

FIG. 14−3: Estimated maximum frequency of lift with two postures (Adapted from Garg and Herrin, 1979).
SOURCE: "Work Practices Guide for Manual Lifting," NIOSH Publication 81−112, National Institute for Occupational Health and Safety, Cincinnati, OH 45226, March 1981.

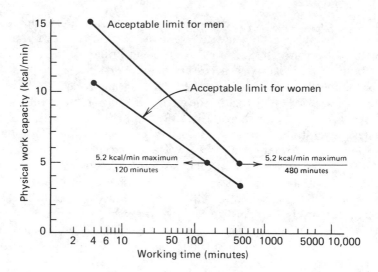

FIG. 14–4: Recommend maximum capacities for continuous work.
SOURCE: "Work Practices Guide for Manual Lifting," NIOSH Publication 81–112, National Institute for Occupational Health and Safety, Cincinnati, OH 45226, March 1981.

personal capabilities and limitations vary from one employee to the next, the frequency of the need for rest should be reviewed by the health and safety officer.

For physical work, one way to measure the effect of repetitive tasks is to calculate the metabolic rate at various loads and frequencies. Figure 14–3 shows the relationship of load and frequency necessary to result in a metabolic rate of 5.2 kcal/min, which is appropriate for men over a full workshift, and for women with working times of less than 2 hours (see Figure 14–4).

B Work Environment Analysis

The work environment includes not only the tools and fixtures with which the work is done, but also such other factors as lighting, heat, humidity, and noise, the so-called comfort indices.

C Furniture and Fixtures

Laboratory furniture should be designed so that occupants can use it with ease and comfort for the entire time they are in the lab. Table 14–1

TABLE 14-1   Standards for Laboratory Furniture

Laboratory Furniture	Standards
Fume Hoods	
Interior	Stainless steel
Width	2.44 m (8 ft)
Height	Vertical with "infinitely" adjustable sashes
Face Velocity	30 ± 6 m/min (100 ± 20 fpm)
Lab Bench	
Height	5.0 cm below worker's elbow (2 in.)
Recording area	30.48 cm wide × 30.48 cm deep (12 in. wide × 12 in. deep)
Chair (adjustable)	
Height (assuming table height = 63.5 cm (25 in.)	38.1−50.8 cm (15−20 in.)
(Assuming table height = 96.52 cm (38 in.)	71.12−83.82 cm (28−33 in.)
Fabric	Vinyl
Seat area	38.1 cm wide × 40.64 cm deep (15 in. wide × 16 in. deep)
Angle between seat pan and backrest (fore/aft adjustability)	105 degrees
Backrest height	17.78−25.4 cm (7−10 in.)
Backrest area	20.32−25.4 cm height × 35.56 cm width (8−10 in. height × 14 in. width) Convexly shaped (50.8 cm radius or 20 in. radius)

Source: C. Mond, et al., "Human Factors in Chemical Containment Laboratory Design," *Am. Ind. Hyg. Assoc. J.*, 48: 10 (October 1987).

presents criteria for the design of ergonomically correct laboratory furniture. Table 14−2 lists criteria for drawers, shelves, and other storage areas in laboratories.

D   Comfort Indices

The amount and nature of the light provided in a laboratory can significantly affect the ease with which work can be accomplished. The amount of

TABLE 14-2   Standards for Design of Storage Areas

Storage Areas	Standard
Drawers and cabinets	Flush to the furniture
	Design for predicted use and projected expansion
	Proper damping and padding for drawers holding glassware
Shelving for storage closets	
Below waist height	Shelf depth should not exceed 45.7 cm (18 in.)
Above shoulder height	Shelf depth should not exceed 30.48 cm (12 in.)
Between waist and shoulder	Shelf depth should not exceed 60.96 cm. (24 in.)
Barrels	Grounded and bonded

Source: C. Mond et al., "Human Factors in Chemical Containment Laboratory Design," *Am. Ind. Hyg. Assoc. J.*, 48:10 (October 1987).

glare, the provision of too much or too little light, and the effects of fluorescent versus incandescent bulbs are all important considerations. Standards for the quantity and nature of light for laboratory environments are presented in Table 14-3.

Criteria for temperature and noise are presented in Table 14-4. The noise level recommended for laboratories is 65 dBA, which is 20 dBA lower than the action level of 85 dBA prescribed by the Occupational Safety and Health Administration (OSHA). This will provide a quiet environment in which work can be performed. If noise levels exceed 85 dBA, certain control measures, as described in the OSHA Noise Standard (29 CFR 1910.95), may be necessary.

The temperature of a laboratory may be of particular importance if persons working in the area are in poor physical condition and/or must use chemical protective clothing. Since protective garments tend to accumulate body heat, persons wearing them will become more uncomfortable more quickly in warm environments. In addition, the use of a respirator may contribute further, adding to heat retention and raising the anxiety level of the user. The use of personal protective clothing in warm and/or humid environments by persons in poor physical condition, or with a low tolerance for heat, is a serious issue. All personnel assigned to wear respirators must be medically approved (see Chapter 19), and heat stress issues should be reviewed by the laboratory safety officer.

TABLE 14-3 Standards for Laboratory Illumination

Environmental Parameter	Standard
LIGHTING	
Intensity	Soft white lighting is preferable
Main lab area	500 lux (50 fc)
Isolation room	500 lux (50 fc)
Benchtop work surface	100-1,000 fc
Work surface under hood	100-1,000 fc
Interior entryway	100 lux
Corridors of main lab	300 lux
Viewing corridors	100-200 lux
Personnel pass-through	200 lux
Changing area	500 lux
Shower stall	200 lux
Restroom/grooming center	200 lux
Luminaire Spacing	2.5 ft from the wall to the center of the luminance
Reflectance	
Ceilings	60-90%
Walls	50-85%
Windows	15-45%
Furniture	30-40%
Floors	15-35%

Source: C. Mond. et al., "Human Factors in Chemical Containment Laboratory Design," *Am. Ind. Hyg. Assoc. J.*, 48:10 (October 1987).

TABLE 14-4 Guidelines for Noise and Temperature in Laboratory Facilities

Noise	
Desirable levels	65 dBA
devices to help achieve goal	Intercoms
	Telephone near entrance
Temperature	70°F: demanding visual and motor tasks
	72°F: secondary tasks
	78°F: showering

Source: C. Mond et al., "Human Factors in Chemical Containment Laboratory Design," *Am. Ind. Hyg. Assoc. J.*, 48:10 (October 1987).

E Analysis of the Worker–Environment Interface

The analysis of the interface between a worker and the environment in which the appointed task must be performed has much to do with making and breaking connections between the person and the environment. In the case of chemical, physical, and biological risks, the goal is to keep the worker separate from the environment. On the other hand, the flow of information to laboratory personnel and the operation of tools, instrumentation, and machines by the personnel should be facilitated.

F Protective Strategies

The nature of the strategies implemented in a laboratory to control or prevent exposure to an agent, or to prevent the release of an agent to the external environment, has much to do with how an employee does a job. Personal protective clothing, if it is worn, must provide for sufficient mobility, tactile sensitivity, and visibility, and yet afford necessary protection against exposure. Issues of claustrophobia or temperature extremes may have to be considered.

Various enclosures, such as specifically designed tissue trimming stations, glove boxes, or downdraft tables, have human factors implications.

One example of this design process is the development of an enclosure for an analytical balance. Not only was the prevention of personnel chemical exposure of primary concern as the enclosure was developed, but human factors issues, such as access, visibility, and range of motion, were also found to be important. The final design incorporated sliding Plexiglas doors for access and visibility. These doors are shut during the actual weighing process of prevent disturbance of the balance. Finally, the dimensions of the balance and the openings reflected a concern for the range of motions the technician executes during the weighing process.

The new enclosure was tested by technicians to determine its suitability and to suggest refinements in its design. The test revealed concerns about visual distortion caused by the Plexiglas, hindrances to manipulative abilities, and the inadequacy of the working area. Redesign of the analytical balance cover addressed these and other issues, and resulted in a safer, more controlled working environment for persons weighing out neat chemical.

G Tools and Instrumentation

The operation of tools and instrumentation in a laboratory environment may acquire a higher degree of risk if extremely toxic or hazardous substances are being handled, or if the use of gloves and personal protective equipment interferes with the ability of the operator. For this reason, careful consideration should be given to the interaction of the operator and the tool, and special accommodation made where advisable. This may include the redesign of buttons and switches for compatibility with gloves, or the implementation of other changes to enhance the reliability and simplicity of operation.

The second aspect of operating laboratory instrumentation is the manner in which information is provided to the operator. Guidelines have been established for illumination and contrast of signs and gauges (Table 14−5), and the issue of visual display terminal (VDT) safety has been given close scrutiny as well.

The environment in which the VDT is used is of concern from a human factors engineering perspective. Many of the complaints voiced by

TABLE 14−5 Standards for Visual Displays

Display Type	Standard	
Warning Lights		
Color selection	Red: danger, warning, fire	
	Yellow: caution	
	Green: go ahead; systems OK	
	Flashing light: extreme danger	
Signs		
Factors to consider	Shape	
	Conspicuousness	
	Width-to-height ratio of 1:6	
	Contrast between characters and	
	background	
Color contrast yielding		
greater visual efficiency	*Characters*	*Background*
	Black	White
	Black	Yellow
	White	Black
	Dark blue	White
	White	Dark red and green
	Dark green and red	White
Dials and Gauges		
Format type	Rectangular	

Source: C. Mond et al., "Human Factors in Chemical Containment Laboratory Design," *Am. Ind. Hyg. Assoc. J.*, 48:10 (October 1987).

VDT users can be addressed by workplace redesign to reduce glare, provide adjustable furniture and seats, and reduce building heat.

III CHECKLIST OF HUMAN FACTORS ISSUES FOR LABORATORIES

A. Work station
   1. Lighting
   2. Range of motion requirements
   3. Furniture design
   4. Convenience of raw materials and waste disposal
B. Tools and instrumentation
   1. Ease of control
   2. Location and visibility of displays
   3. Level of physical effort required
C. Environmental considerations
   1. Temperature
   2. Relative humidity
   3. Noise levels
   4. Lighting
D. Personal protective equipment
   1. Mobility
   2. Claustrophobia and other psychological issues
   3. Heat buildup
   4. Visibility
   5. Dexterity and tactile sensitivity
E. Emergency equipment
   1. Availability
   2. Location
   3. Ease of donning
   4. Appropriateness to hazard
F. Floor plan evaluation
   1. Traffic pattern
   2. Contamination containment
   3. Hazards of slipping, tripping, or falling
   4. Accessibility to showers and emergency equipment
G. Lifting and manual handling of materials
   1. Weight
   2. Location
   3. Frequency
   4. Duration

IV CHEMICAL HYGIENE PLAN CONSIDERATIONS

Despite the absence of specific requirements for human factors to be addressed in the hygiene plan, it is recognized that both safety and efficiency depend heavily on a properly designed laboratory. The development and implementation of such a design should be a priority of all laboratory managers.

RESOURCES

ANSI Standard RP-7−1979, "American National Standard Practice for Industrial Lighting." New York: Illuminating Engineering Society, 1979.

Boyce, P. R., *Human Factors in Lighting*. New York: Macmillan, 1981.

Cakir, A., D. J. Hart, and T. F. M. Stewart, *Visual Display Terminals*. New York: Wiley, 1980.

Damon, A., H. W. Stoudt, and R. A. McFarland, *The Human Body in Equipment Design*. Cambridge, MA, Harvard University Press, 1966.

Harless, J., "Components in the Design of a Hazardous Chemicals Handling Facility," in *Health and Safety for Toxicity Testing*. D. B. Walters, and C. W. Jameson, Eds.; Boston: Butterworth, 1984, pp. 45−71.

Huchingson, R. D., *New Horizons for Human Factors in Design*. New York: McGraw-Hill, 1981.

Konz, S., *Work Design*. Columbus, OH: Grid Publishing, 1979.

McCormick, E. J., *Human Factors in Engineering and Design*, 4th ed. New York: McGraw-Hill, 1976.

Mond, C., et al., "Human Factors in Chemical Containment Laboratory Design," *Am. Ind. Hyg. Assoc. J.*, 48:10 (October 1987).

National Institute for Occupational Safety and Health, "Work Practices Guide for Manual Lifting," NIOSH Publication 81−122, Cincinnati, OH.

Poulton, E. C., *Environment and Human Efficiency*. Springfield, IL: Charles C. Thomas, 1970.

Van Cott, H. P., and R. G. Kinkade (Eds.), *Human Engineering Guide to Equipment Design* (rev. ed.). Washington, DC: Government Printing Office, 1972.

Woodson, W. E., *Human Factors Design Handbook*. New York: McGraw-Hill, 1981.

CHAPTER 15

SAFETY SHOWERS AND EYEWASH STATIONS

I INTRODUCTION

Prevention of employee exposure to hazardous materials is at the core of laboratory health and safety programs. Although prevention is important in any safety program, accident anticipation and emergency response are also critical components of a comprehensive program. A facility must establish means of responding to accidental eye and body exposure to the chemicals in use.

Basic guidelines and regulations for eyewash stations and safety showers have evolved in recent years. This chapter discusses these regulations and the use of eyewash stations and safety showers in a laboratory. It also addresses the different types of equipment available, as well as the location and maintenance of these safety appliances.

II SELECTION, LOCATION, AND INSTALLATION OF EQUIPMENT

A laboratory must locate safety showers and eyewash stations throughout the facility. The equipment should be located close to where staff may be using potentially hazardous materials. The equipment used and its placement must meet local, state, and federal regulations.

Several different types of equipment comply with federal, state, and local regulations. A facility's health and safety officer should assess the specific needs of each laboratory in relation to hazards, regulations, and building design, and then recommend the best equipment for each laboratory within the facility. The following sections describe criteria that should be used in this assessment.

A Federal Regulations

Current standards of the Occupational Safety and Health Administration (OSHA) discuss eyewash stations and safety showers only in general terms. OSHA suggests that personal protective equipment in a laboratory include an easily accessible drench-type safety shower and an eyewash station within the work area. OSHA has also addressed basic recommendations for maintenance, labeling, and training in the use of this equipment. Table 15−1 summarizes OSHA standards for eyewash stations and safety showers.

The American National Standard Institute (ANSI) has published a more detailed voluntary industry standard, which covers physical features, location, and maintenance for this equipment. ANSI document Z358. 1−1981 presents design and performance requirements for six different types of emergency eyewash and shower equipment. Table 15−2 reviews ANSI Z358.1−1981.

TABLE 15−1 OSHA Standards

29 CFR 1910.151(c)

Where the eyes or body of any person may be exposed to injurious corrosive materials, suitable facilities for quick drenching or flushing of the eyes and body shall be provided within the work area for immediate emergency use.

29 CFR 1910.94(d) (9)(vii)

Near each tank containing a liquid that may burn, irritate, or otherwise be harmful to the skin if splashed upon the worker's body, there shall be a supply of clean, cold water. The water pipe (carrying a pressure not exceeding 25 lb) shall be provided with a quick-opening valve and at least 48 in. of hose not smaller than three-fourths in., so that no time may be lost in washing off liquids from the skin or clothing. Alternatively, deluge showers and eye flushes shall be provided in cases where harmful chemicals may be splashed on parts of the body.

TABLE 15-2 ANSI Z358.1-1981 Standards for Emergency Eyewash and Shower Equipment

Type of Equipment	Physical Specifications	Laboratory Location	Maintenance	Training
Emergency showers	Water column between 82 and 96 in. with 20-in. minimum diameter column at 60 in. above surface; should deliver 30 gal/min (gpm); enclosures, if used, require minimum 34-in. unobstructed diameter	Accessible within 10 seconds and not more than 100 ft from hazard	Activated weekly to flush lines and verify operation	Required for all employees who might be exposed to a chemical splash
Plumbed and self-contained eyewashes	Flow rate of 0.4 gpm for 15 minutes required; water nozzles 33–45 in. above floor	Accessible within 10 seconds and not more than 100 ft from hazard	Plumbed units activated weekly to flush lines and verify operation; self-contained units treated in accordance with manufacturer's instructions	Required for all employees who might be exposed to a chemical splash
Personal eyewashes	Not addressed	Not specified, but recommended to be placed in immediate vicinity of potentially hazardous areas	Tested, refilled, disposed, and maintained in accordance with manufacturer's instructions; activated weekly to flush lines and verify operation	Required for all employees who might be exposed to a chemical splash

TABLE 5.2 (Continued)

Eye/face washes	Flow rate of 3.0 gpm for 15 minutes required; water nozzles 33–45 in. above floor	Accessible within 10 seconds and not more than 100 ft from hazard	Activated weekly to flush lines and verify operation	Required for all employees who might be exposed to a chemical splash
Hand-held drench hoses	Flow rate of 3.0 gpm required	Accessible within 10 seconds and not more than 100 ft from hazard	Activated weekly to flush lines and verify operation	Required for all employees who might be exposed to a chemical splash
Combination units	Must meet physical requirements of component parts	Accessible within 10 seconds and not more than 100 ft from hazard	Activated weekly to flush lines and verify operation	Required for all employees who might be exposed to a chemical splash

B Types of Eyewash Stations and Safety Showers

Eyewash stations and safety showers are available in several different types. ANSI has grouped the eyewash and shower appliances that may be used in a laboratory into the following categories:

- *Emergency shower*: "a unit that enables the user to have water cascading over the entire body."
- *Plumbed and self-contained eyewash*: a plumbed unit is "an eyewash unit permanently connected to a source of potable water"; a self-contained eyewash is one "that is not permanently installed and must be refilled or replaced after use." Self-contained eyewashes must have at least a 15-minute water supply.
- *Personal eyewash*: "a supplementary eyewash that supports plumbed units, self-contained units, or both, by delivering immediate flushing for less than 15 minutes."
- *Eye–face wash*: "a device used to irrigate and flush both the face and the eyes."
- *Hand-held drench hose*: "a flexible hose connected to a water supply and used to irrigate and flush eyes, face, and body areas."
- *Combination unit*: "a unit combining a shower with an eyewash or eye/face wash, or with a drench hose, or with both, into one common assembly."

C Location

The effectiveness of any eyewash station or safety shower depends on its accessibility. The first 15 seconds after an injury occurs are critical, so a laboratory should place the emergency shower and/or eyewash station close to the hazard site. ANSI recommends accessibility within 10 seconds and placement at distance no greater than 100 ft. ANSI also suggests that the laboratory place eyewash fountains and safety showers near the laboratory entrance. Selection of a location should also be a function of traffic patterns, the specific contaminants, the number of personnel performing hazardous operations, the specific hazards to be protected against, and whether protective equipment is worn. Also, workers should have easy access to the eyewash station or safety shower, without intervening partitions or obstructions. The laboratory should pay particular attention to the proximity to electrical outlets and the length of any shower pull chains in a walkway. Also, the laboratory should store a blanket close to either an eyewash or safety shower to prevent shock and provide privacy.

D Water Supply

To satisfy general requirements, laboratory personnel should flush contaminants with "copious" amounts of water. An eyewash station or safety shower should, therefore, deliver a slow stream of water for at least 15 minutes. A slow or spent stream of water is preferable, since high pressure in an eyewash may drive particulate hazards into the eyes. ANSI Z358.1−1981 recommends a pressure of 30 psi in an eyewash and provides other guidelines for ensuring water flow (Table 15−2). Laboratory workers should be able to operate the eyewash with push-to-operate actuation valves that will remain open until manually closed. Such valves free the hands and allow the injured person to hold back the eyelids for a thorough flushing.

A laboratory should provide only potable water in its safety stations, and the temperature of the water should be kept within a comfortable range (60−95°F) to prevent shock and to encourage usage. Temperatures above 100°F are not desirable because they increase circulation and may increase absorption of the chemical. Water temperatures above 120°F may cause first-degree burns. Finally, a laboratory should see that any outdoor showers and eyewashes provide tempered water, which may require the addition of a heated holding tank.

Where self-contained eyewash units are used, a program of frequent water replacement must be adopted. Harmful microorganisms have been shown to grow in these units, and introduction of contaminated water into the eye can cause infection and, in severe cases, loss of sight.

III OPERATION

A Maintenance

A laboratory should activate its plumbed eyewashes and safety showers weekly to flush the lines and to permit observation of proper pressurization levels. It should also conduct a documented inspection of water pressure on a montly basis, testing its portable units and checking that the fill level is in accord with the manufacturer's instructions.

B Use and Training

While it is important that a laboratory supply proper eyewash stations and safety showers, it is equally important that the laboratory train personnel in their proper use.

The laboratory should document emergency procedures in writing and should properly label safety equipment. Also, it should see that its personnel are familiar with the controls and operating devices, as well as the procedures involved in assisting an injured person. Very often an injured person cannot flush his or her own eyes, and two people are needed — one to hold open the victim's eyes and the other to restrain the victim, who may be in pain. The laboratory should introduce its personnel to the proper methods involved in such an emergency. Also, it should clearly label its safety showers and eyewash stations, like all safety equipment.

Laboratory personnel should never use neutralizing chemicals, and boric acid should never be used when a chemical has been introduced to the body. Such actions may actually increase the injury.

IV CHEMICAL HYGIENE PLAN CONSIDERATIONS

The chemical hygiene plan should include the laboratory's procedure for responding to chemical exposure accidents. Therefore, it should address the location and proper use of eyewash and safety shower equipment in the facility. The chemical hygiene plan should also describe the maintenance, inspection, and testing program in place for this equipment.

V MANUFACTURERS AND SUPPLIERS

The Thomas 1987 Register lists the following manufacturers and suppliers of eyewash and safety shower equipment:

Bradley Corporation, Menomonee Falls, Wisconsin
Eastoo Industrial Safety Corporation, New York, New York
Guardian Equipment, Chicago, Illinois
Haws Drinking Faucet Company, Berkeley, California
Intest, Inc., Newport News, Virginia
Ogontz Corporation, Willow Grove, Pennsylvania
Safety Equipment Company, Tallahassee, Florida
Safety Services, Inc., Kalamazoo, Michigan
Sargent-Sowell, Inc., Grand Prairie, Texas
Sipco Products, Inc., Peoria, Illinois
Speakman Company, Wilmington, Delaware
Water Saver Faucet Company, Chicago, Illinois

Western Drinking Fountains, Emergency Equipment Division, Glen Riddle, Pennsylvania

RESOURCES

American National Standard for Emergency Eyewash and Shower Equipment, ANSI 2358.1–1981. New York: American National Standards Institute, 1981.

Jonathan, R.D., "Selection and Use of Eyewash Fountains and Emergency Showers," *Chem. Eng.*, pp. 147–150, Sept. 15, 1975.

Occupational Safety and Health Administration. Code of Federal Regulations, Title 29 Parts, 1900–1910. Washington, DC.

Russell, B., "Tempered Water for Safety Showers and Eye Baths," *Chem. Eng.*, pp. 95–97, Nov. 24, 1983.

Thomas Register of American Manufacturers and Thomas Register Catalog File, 77th ed., New York: Thomas Publishing, 1987.

Walters, D.B., Stricoff, R.S., and Ashley, L.E., "The Selection of Eyewash Stations for Laboratory Use." (unpublished data).

CHAPTER 16

CHEMICAL EXPOSURE EVALUATION

I INTRODUCTION

The purpose of chemical monitoring is to ensure that exposures of personnel to hazardous materials are within acceptable limits. Exposure monitoring consists of identifying and evaluating sources of exposure, then measuring chemical concentrations. Exposure monitoring requires knowledge of the monitoring methods and of where exposures are likely to occur. Thus, the implementation of an exposure monitoring program can be complex.

II TYPICAL CHEMICAL EXPOSURES IN LABORATORIES

Operations that are typically conducted in laboratories may create a variety of exposure risks. Often, there are uncommon chemicals in use which have unknown physical, chemical, and toxicological properties.

Commonly used laboratory reagents may also present exposure risks that require evaluation. Chemicals requiring monitoring may be present in several physical states:

Gases. Chemicals may exist as a gas under normal laboratory conditions. The primary route of exposure to these chemicals is inhalation.

Vapors. Vapors exist in the gaseous state in equilibrium with a liquid or solid chemical, or when such a chemical is heated. The primary route of exposure to vapors is inhalation.

Mists. Mists are aerosolized droplets of liquid chemical usually created by some mechanical action in a laboratory procedure. The primary routes of exposure to mists are inhalation and skin absorption.

Fumes. Fumes are aerosolized solid particulates created by condensation of a solid that has been heated to form a vapor and then cooled. The primary route of exposure to fumes is inhalation.

Dusts. Dusts are generally created by a mechanical action on a solid. They are dispersed into the air by the mechanical action. The particle size of the dust determines the extent of dispersion, and whether it can be inhaled. It also determines where the dust will be deposited in the respiratory tract. The primary routes of exposure to dust are inhalation and ingestion. However, chemically contaminated dust, in some cases, may create a skin absorption problem.

III   IDENTIFICATION OF MONITORING PRIORITIES

Each laboratory should develop a sampling strategy plan by observing the processes and operations it uses. In theory, a laboratory could measure the exposures resulting from all its operations; however, this is usually impractical and unnecessary. Therefore, to implement a sampling strategy, the laboratory must specify monitoring priorities for situations in which employees may be highly prone to chemical exposure.

The sampling logic diagram in Figure 16-1 offers one approach for determining whether chemical sampling is required. A planning committee should conduct a preliminary walk-through survey of the laboratory to identify priority operations and chemicals that present the greatest health risk potential. After identification of these operations, chemicals, and the personnel potentially exposed, the committee should structure a sampling strategy plan.

IV   DEVELOPING A SAMPLING STRATEGY PLAN

The sampling strategy plan should be based on good industrial hygiene practice and on the employees' exposure potential, the frequency of exposure, and the particular hazards of the chemicals to which employees may be exposed. In some cases, the Occupational Safety and Health Administration (OSHA) may require the use of a particular sampling

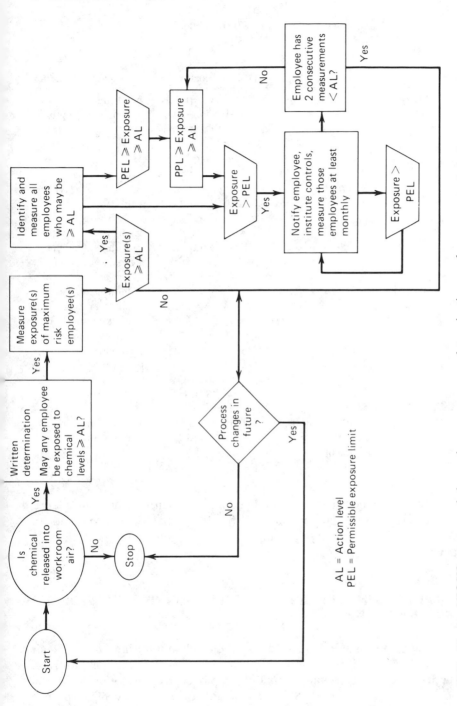

FIG. 16–1: NIOSH-recommended employee exposure determination and measurement strategy.

TABLE 16−1 Example of a Sampling Strategy Plan

Agent	Hazard	Observed or Suspected Exposure	Sampling Frequency
Formaldehyde	Potential carcinogen	OSHA exposure standard of 1 ppm 8-hour time weighted average	Monthly if > 1 ppm, otherwise annually
Xylene	Toxic: threshold limit value (TLV) = 100 ppm	Routine exposure up to 100 ppm	Annually
Diethyl ether	Toxic. TLV = 400 ppm	< 50 ppm intermittently	Exposure — annually with regular use Peroxides — before use and every 6 months; never leave an open container in storage more than 2 years
Carbon disulfide	Toxic: TLV = 10 ppm	Routine exposure within the TLV	6 months
Ethylene oxide	TLV = 1 ppm Action level = 0.5 ppm	Above action level and less than TLV	Every 6 months, as required by OSHA

strategy for a specifically regulated chemical. When the sampling plan is unspecified, one similar to that of Table 16−1 should be developed for each operation selected for sampling in the preliminary walk-through survey.

V FACTORS INFLUENCING THE SELECTION OF A MONITORING TECHNIQUE

There are four major categories of chemical monitoring: personal air monitoring, area air monitoring, wipe sampling, and biological monitoring. The first three categories are commonly employed in laboratories; biological monitoring is rarely used. Selection of the correct monitoring approach depends on assessments made when exposures are initially identified. Laboratories selecting a monitoring approach should answer the following questions:

1. What is the frequency and duration of the exposure, and how many employees are exposed?
2. What are the possible routes of exposure and the expected airborne concentrations of the chemical?
3. What are the physical, chemical, and toxicological properties of the agent to be monitored?
4. What sampling method should be employed?
5. What is a representative sample, and is there a need to measure peak exposures?
6. What environmental conditions could impact the sampling methods (e.g., temperature, humidity, air currents, other operations in the area), and what are the physical and time constraints on sampling the operation?
7. What range of exposure levels is possible?
8. Is there a reliable analytical procedure? What potential interferences exist, and what are the properties (detection limit, range, precision, and accuracy) of the method?

VI TYPES OF MONITORING METHOD

A Personal Sampling

Personal sampling involves collecting a sample and using a sampler placed as close as possible to the worker's breathing zone, with the device mounted on the worker so that it moves through the laboratory as the worker does. This is the best method for estimating the worker's actual chemical exposure. Thus, as a rule, if the goal of sampling is to measure a worker's exposure, the laboratory should use personal sampling. There are three ways in which personal monitoring is performed:

Full-period sampling,
Partial-period consecutive sampling, and
Grab sampling.

A laboratory employs full-period single monitoring to sample the concentrations of a chemical over a full shift. It is a preferred method because it provides a measurement of an employee's time-weighted average daily exposure. The full-period sample may actually be a composite of several samples when the laboratory wishes to evaluate peak exposures within a work shift, or when the collection capacity of the monitoring method precludes a single, longer sample.

A laboratory will employ partial-period monitoring when it expects a worker's daily exposure to remain reasonably constant over an entire shift. In such cases, a partial-shift sample can provide a representative estimation of a worker's daily exposure, as long as the laboratory collects enough sample to meet the detection limit requirements of the analytical method.

Grab sampling provides a point-in-time "snapshot" of the airborne concentration of a chemical. This method is extremely useful in identifying

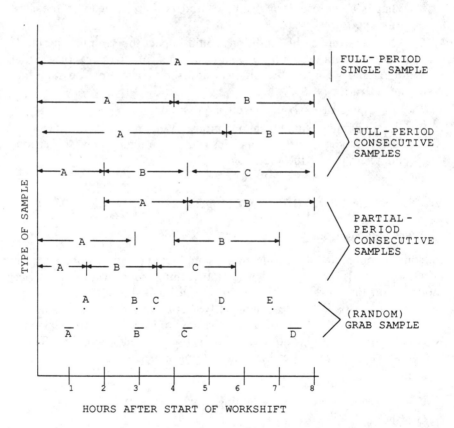

FIG. 16—2: Data analysis procedures available for sampling. Reference chart of types of exposure measurement that could be taken for an 8-hour average exposure standard.
SOURCE: N.A. Leidel, et al., *Occupational Exposure Sampling Strategy Manual*, NIOSH, January 1977.

task-related exposures and subsequently determining whether exposure control methods are still working as needed.

Figure 16–2 presents a pictorial representation of these methods, which are often used in combination to evaluate an employee's chemical exposures. It is important to collect a large enough sample to assure that exposures are accurately determined within statistical confidence limits. Typically, a 95% confidence limit is considered satisfactory.

B   Area Air Monitoring

In area monitoring sampling devices are placed around the source of exposure. This is a very useful technique for quantifying the emissions from specific sources, and differs from personal sampling in that area sampling measures the exposure level at a fixed point. Because area samples do not reflect the effects of an individual's movements within the workplace, this approach can result in significant underestimation or overestimation when used to estimate personal exposures.

C   Wipe Monitoring

Wipe monitoring is useful for investigating dispersion of a condensed phase material from an exposure source. The technique involves using surface area wipes to determine the presence and quantity of a particular chemical at various positions of interest throughout the work area. This method can also be used to determine whether a worker's protective equipment is free of contamination.

D   Biological Monitoring

Biological monitoring involves the measurement of a chemical or its metabolites in body fluids (or, in some cases, hair or nails). This method directly measures a worker's body burden of a specific chemical, but techniques for performing biological monitoring are limited to relatively few chemicals. Chapter 17 provides more information on biological monitoring.

VII   EQUIPMENT AND INSTRUMENTATION

A   Active Sampling

Numerous sampling methods are available for monitoring a wide variety of chemical exposures. Some involve sampling with an air pump, which draws contaminated air through a sampling train at a constant flow rate.

It is very important that the investigator select the correct sampling medium and calibrate the air pump flow rate both before and after monitoring. The next step is to calculate the volume of air sampled and, after the quantity of analyte collected on the sampling medium has been measured, the concentration of analyte in the air (e.g., milligrams of contaminant found per cubic meter of air drawn through the sampler).

B Passive Sampling

Passive dosimeters collect gaseous contaminants in the air by diffusing the analyte through a membrane onto a sorbent. Passive monitoring devices do not require any calibration before or after use, and they have the advantage of being easily transportable and simple to use. Numerous passive dosimetry products that can measure a variety of specific gases are available on the market today.

Section XI lists a number of sampling equipment manufacturers and distributors.

C Grab Sampling

Grab sampling is used in identifying sources of exposure and in "screening", to find areas of highest concentration. Direct-reading instruments, or instantaneous sampling devices, are best suited for this purpose. Examples of this type of equipment include colorimetric tubes, which indicate analyte concentrations by the length of a stain on a solid sorbent, and evacuated cylinders, which are punctured to capture an air sample. This air-sampling technique is generally not useful to evaluate personal exposure levels; however, in combination with other techniques, it can be very useful.

D Wipe Sampling

Wipe sampling involves using an appropriate medium to wipe a given surface area. The area wiped must be the same size in area throughout the investigation so that the data will be comparable with other wipe samples. The wipe medium must also be conducive to the analytical method.

VIII COMMONLY ENCOUNTERED SAMPLING PROBLEMS

Laboratories may encounter a number of problems in making chemical exposure assessments. These include, but are not limited to, the following:

sampling equipment calibration errors,
sample contamination,
varying environmental conditions,
lack of sample homogeneity,
absorption of analyte onto sample container walls,
use of improper sampling medium,
incomplete elution of analyte from sampling medium,
channeling of analyte on the collection medium,
degradation of analyte prior to analyses,
mechanical defects in sampling equipment,
partial vapor pressure effects of gases,
reactivity of the analyte with sampling medium,
volumetric errors and sampling rate errors,
temperature and pressure effects during sampling,
analytical errors, and
calculation errors.

Discussion of the variety of precautions that should be taken to avoid these pitfalls would require an extensive document devoted solely to the subject of industrial hygiene sampling. However, the person conducting the sampling should be sure to consider these problems for the specific chemical monitoring technique being employed. Typically, methods published by the National Institute for Occupational Safety and Health (NIOSH) and OSHA give information on common problems, (e.g. appropriate ranges, interferences). It is best to use the services of a professional industrial hygienist to design detailed sampling and analysis procedures.

IX EVALUATION OF SAMPLING RESULTS

Once the sampling has been completed, several steps should be taken, depending on the outcome of the testing. If the sampling results indicate exposures within acceptable limits, and the results are representative of employee exposures, the results should be reported to the affected employees and documented and filed for future reference. Remember that OSHA regulations require the retention of exposure evaluation records (see Chapter 25, Regulations).

If the sampling results indicate exposures above acceptable levels, the laboratory should implement both immediate and long-term steps. In cases of highly toxic compounds, the laboratory should consider shutting down the operation until exposure-reducing controls can be implemented. In situations involving compounds of a less toxic nature, the laboratory should implement personal protective and engineering controls of various types to provide both immediate and long-term protection to the employees. The laboratory should inform its employees of all sampling results and of steps being taken or planned to reduce exposures. Finally, after implementing permanent controls, the laboratory should conduct further sampling to verify their effectiveness.

X   CHEMICAL HYGIENE PLAN CONSIDERATIONS

The chemical hygiene plan should describe the laboratory's overall strategy for monitoring worker exposures. Some laboratories identify high risk locations and operations, and monitor these periodically. Other laboratories rotate sampling through the facility so that over a multiyear period all locations will be monitored. Still other laboratories do no routine sampling, instead monitoring when a new project raises questions about worker exposures.

The approach that is right for any specific laboratory depends on the nature of the work performed, the control systems in place, and the amount of historical data available. However, the approach should be described, and the rationale for its use explained, in the hygiene plan.

The methods to be used for sampling and analysis, or the way in which they will be selected, should be described. Also, the way in which data will be reported and stored should be described.

XI   REFERENCES AND VENDORS

A number of useful references on exposure sampling are available to assist laboratory staff in developing the proper strategy, sampling technique, and analytical methods to assess exposures properly. Both NIOSH and OSHA have documented chemical monitoring methods for common laboratory chemicals. Authorities have validated these sampling methods and accepted them for practice.

It is also important that a laboratory interpret data correctly. NIOSH, OSHA, the American Conference of Governmental Industrial Hygienists (ACGIH), and the American Industrial Hygiene Association (AIHA) have all published guidelines for assessing chemical exposures in the workplace. Levels identified as acceptable exposure limits for specific chemicals often vary among these sources, and one prudent approach is to use the lowest published exposure standard as an exposure limit (i.e., NIOSH, OSHA, ACGIH). It is just as important to use a good analytical laboratory as it is to use the proper sampling method. For this reason, one should opt for an AIHA-accredited laboratory to analyze samples collected in the field whenever possible.

Table 16−2 lists many of the vendors of chemical monitoring equipment that can assist in implementing a chemical monitoring program in the laboratory.

TABLE 16–2   Sources of Monitoring Equipment

Calibration Gases and Equipment

Air Engineers, Inc., Safety & Health Division	Interstate Safety & Supply, Inc.
American Bristol Industries, Inc.	Jones Safety Supply, Inc.
Ashland Chemical Company	Kurz Instruments, Inc.
Bacharach, Inc.	Lifecom Safety Service & Supply Company
Briggs Weaver, Inc.	Lumidor Safety Products
Calibrated Instruments, Inc.	Mast Development Company
Carey Machinery & Supply Company, Inc.	Matheson Gas Products, Inc.
CEA Instruments, Inc.	MSA
Chapin Ashuelot Medical & Safety Supply	National Draeger, Inc.
Continental Safety Equipment, Inc.	National Mine Service Corporation
CSE Corporation	Newark Glove & Safety Equipment Company, Inc.
Day Star Corporation	Pendergast Safety Equipment
Detcon, Inc.	Precision Flow Devices
Digicolor	Pro Am Safety
Direct Safety Company	ProTech Safety Equipment, Inc.
Dynamation, Inc.	Protective Equipment, Inc.
Eastco Industrial Safety Corporation	Raeco, Inc.
ECI United Safety, Inc.	Reis Equipment Company
Environmental Compliance Corporation	Safety Services, Inc.
Gas Tech, Inc.	Safety Supply Canada
GC Industries, Inc.	Scott Specialty Gases
GT Safety Equipment, Inc.	Sensidyne, Inc.
IMR Corporation	Sierra Monitor Corporation
Industrial Products Company	SKC West, Inc.
Industrial Protective Equipment Supply Company	Standard Safety Equipment Company
International Ecology Systems Corporation	Tackaberry Company

Texas Analytical Controls, Inc.
Thermo Environmental Instruments
Tierney Safety Products
Tracor Atlas, Inc.
Vallen Safety Supply Company
VICI Metronics
Wolsk Alarms, Ltd.

Carbon Monoxide Monitors and Detectors

Acme Engineering Products, Inc.
Advanced Chemical Sensors Company
M. Clifford Agress, PE
Air Engineers, Inc., Safety & Health Division
American Bristol Industries, Inc.
Andersen Samplers, Inc.
Arbill, Inc.
Bacharach, Inc.
Baseline Industries
Biotrak, Inc.
Briggs Weaver, Inc.
Butler National Corporation
Calibrated Instruments, Inc.
Carey Machinery & Supply Company, Inc.
CEA Instruments, Inc.
Chapin Ashuelot Medical & Safety Supply
Chestec, Inc.
Conney Safety Products Company
Continental Safety Equipment, Inc.
Control Instruments Corporation

Critical Services
CSE Corporation
Day Star Corporation
Delaware Valley Safeguard
Devco Engineering, Inc.
Direct Safety Company
Dynamation, Inc.
Dynatron, Inc.
Eagle Air Systems
Eastco Industrial Safety Corporation
ECI United Safety, Inc.
Engwald Corporation
Emmet Corporation
Enterra Instrumentation Technologies, Inc.
Foxboro Company
Gas Tech, Inc.
GfG Gas Electronics, Inc.
GT Safety Equipment, Inc.
Halprin Supply Company
Hazco, Inc.
Health Consultants, Inc.
High Pressure Equipment, Inc.
Horiba Instruments, Inc.
Hub Safety Equipment
Industrial Analytical Laboratory, Inc.
Industrial Products Company
Industrial Protective Equipment Supply Company
Industrial Safety Products, Inc.
Industrial Scientific Corporation
International Sensor Technology

TABLE 16–2   (Cont.)

Interscan Corporation	Raeco, Inc.
Interstate Industrial Supply	Reis Equipment, Inc.
Interstate Safety & Supply, Inc.	Rockford Medical & Safety Company
Jones Safety Supply, Inc.	Roxan, Inc.
Kanton Air Products Corporation	Rubin Brothers
Laboratory Safety Supply Company	Safety Services, Inc.
Lifecom Safety Service & Supply Company	Scott Aviation
Lumidor Safety Products	Sensidyne, Inc.
Macurco, Inc.	Sheridan Safety Supply, Inc.
Mateson Chemical Corporation	Sieger Gasalarm
Matheson Gas Products, Inc.	Sierra Monitor Corporation
MC Products	SKC West, Inc.
MDA Scientific, Inc.	Standard Marketing International, Inc.
Metrosonics, Inc.	Standard Safety Equipment Company
Mine Safety Appliance Company	Sunshine Instruments
MSA	Syracuse Safety Services, Inc.
National Draeger, Inc.	Tackaberry Company
National Mine Service Company	Thermo Environmental Instruments
Neotronics	Trace Analytical
Neutronics, Inc.	Tracor Atlas, Inc.
Newark Glove & Safety Equipment Company, Inc.	Trusafe, Inc.
Frank Niemi Products, Inc.	US Industrial Products Company, Inc.
Pedly & Knowles & Company	US Safety, Cecso Service Company
Pendergast Safety Equipment	Vallen Safety Supply Company
Pro Am Safety	Ward International
ProTech Safety Equipment, Inc.	Wise El Santo Company
Racal Airstream, Inc.	Wolsk Alarms Ltd.

Detector Tubes

Advanced Chemical Sensors Company
Air Engineers, Inc., Safety & Health Division
American Bristol Industries, Inc.
Arbill, Inc.
Automation Products, Inc.
Bacharach, Inc.
BGI, Inc.
Carey Machinery & Supply Company, Inc.
Chapin Ashuelot Medical & Safety Supply
Chemrox, Inc.
Continental Safety Equipment, Inc.
Day Star Corporation
Delaware Valley Safeguard
Detector Electronics Corporation
ECI United Safety, Inc.
Edcor Safety
Emmet Corporation
Environmental Compliance Corporation
Fire House
Foxboro Company
GT Safety Equipment, Inc.
Hazco, Inc.
Health Consultants, Inc.
Industrial Analytical Laboratory, Inc.
Industrial Products Company
Interstate Safety & Supply, Inc.
Jones Safety Supply, Inc.
Lifecom Safety Service & Supply Company

Matheson Gas Products, Inc.
Mine Safety Appliance Company
MSA
National Draeger, Inc.
National Mine Service Company
PCP, Inc.
Pedly & Knowles & Company
Pendergast Safety Equipment
ProTech Safety Equipment, Inc.
Protective Equipment, Inc.
Raeco, Inc.
Reis Equipment Company
Roxan, Inc.
Safety Services, Inc.
Safety Supply Canada
Scott Specialty Gases
Sensidyne, Inc.
Sheridan Safety Supply, Inc.
AJ Sipin Company, Inc.
SKC, Inc.
SKC West, Inc.
Standard Safety Equipment Company
Syracuse Safety Services, Inc.
Tackaberry Company
Tierney Safety Products
Vallen Safety Supply Company

Formaldehyde Monitors and Detectors

Advanced Chemical Sensors Company

TABLE 16–2 (Cont.)

Aetna Technical Services, Inc.	HNU Systems, Inc.
Air Engineers, Inc., Safety & Health Division	Holland Safety Supply
Air Quality Research	Industrial Analytical Laboratory, Inc.
American Medical Laboratories Inc., Industrial Hygiene Division	Industrial Products Company
Anacon	Industrial Protective Equipment Supply Company
Andersen Samplers, Inc.	International Ecology Systems Corporation
Arbill, Inc.	International Sensor Technology
Assay Technology, Inc.	Interstate Safety & Supply, Inc.
Bacharach, Inc.	Laboratory Safety Supply Company
Baseline Industries	LaMotte Chemical
Butler National Corporation	Macurco, Inc.
Carey Machinery & Supply Company, Inc.	Mateson Chemical Corporation
CEA Instruments, Inc.	Matheson Gas Products, Inc.
Continental Safety Equipment, Inc.	MDA Scientific, Inc.
Day Star Corporation	National Draeger, Inc.
Delaware Valley Safeguard	National Mine Service Company
Devco Engineering, Inc.	PCP, Inc.
Direct Safety Company	Delta Thermographics, Inc.
Eastco Industrial Safety Corporation	Dyn Optics
ECI United Safety, Inc.	Epic, Inc.
Edcor Safety	Fox Valley Systems, Inc.
Emmet Corporation	Foxboro Company
Environmental Compliances Corporation	Gas Tech, Inc.
Foxboro Corporation	Horiba Instruments, Inc.
GT Safety Equipment, Inc.	Ikegami Electronics USA, Inc.
Hager Laboratory, Inc.	International Light, Inc.
Hazco, Inc.	Interstate Safety & Supply, Inc.

MC Products
MSA
Neotronics
Pacer Industries, Inc.
Raeco, Inc.
Teledyne Analytical Instruments

Personal Monitors

Advanced Chemical Sensors Company
Air Engineers, Inc., Safety & Health Division
Air Quality Research
American Gas & Chemical Company Ltd.
American Medical Laboratories, Inc., Industrial Hygiene Division
Anacon
Andersen Samplers, Inc.
Arbill, Inc.
Asbestos Control Technology, Inc.
Assay Technology, Inc.
Audio Medical, Inc.
Bacharach, Inc.
Baird Corporation
BGI, Inc.
Briggs Weaver, Inc.
Butler National Corporation
Carey Machinery & Supply Company, Inc.
Chapin Ashuelot Medical & Safety Supply
Chemrox, Inc.
Continent Safety Equipment, Inc.

Critical Services
Day Star Corporation
Delaware Valley Safeguard
Devco Engineering, Inc.
Direct Safety Company
Dosimeter Corporation of America
Du Pont de Nemours & Company, Inc.
Dynamation, Inc.
Eastco Industrial Safety Corporation
ECI United Safety, Inc.
Emmet Corporation
Environmental Compliance Corporation
Environmental Safety Products, Inc.
ESA Laboratories, Inc.
Gabriel Environmental Energy
GasTech, Inc.
GC Industries, Inc.
GfG Gas Electronics, Inc.
GMD Systems, Inc.
Grace Industries, Inc.
GT-Safety Equipment, Inc.
Hager Laboratory, Inc.
Halprin Supply Company
Hazco, Inc.
Health Consultants, Inc.
Honba Instruments, Inc.
ICN Dosimetry Services
Ikegami Electronics USA, Inc.
Impact Hearing Conservation, Inc.

TABLE 16–2 (Cont.)

Industrial Hygiene Specialties Company
Industrial Products Company
Industrial Protective Equipment Supply Company
Industrial Safety Products, Inc.
Industrial Scientific Corporation
International Sensor Technology
Interscan Corporation
Interstate Safety & Supply, Inc.
Pendergast Safety Equipment
Photovac, Inc.
ProTech Safety Equipment, Inc.
Raeco, Inc.
Reis Equipment Company
Roxan, Inc.
Safety Services, Inc.
Safety Supply Canada
Sensidyne, Inc.
Sentex Sensing Technology, Inc.
Sheridan Safety Supply, Inc.
Sierra Monitor Corporation
SKC, Inc.
SKC West, Inc.
Standard Safety Equipment Company
Syracuse Safety Services, Inc.
Tackaberry Company
Thomas Scientific
3M Company Occupational Health & Safety Products
Tierney Safety Products

US Industrial Products Company, Inc.
Vallen Safety Supply Company
Wise El Santo Company

Hydrocarbon Detectors and Analyzers

Andersen Samplers, Inc.
Bacharach, Inc.
CEA Instruments, Inc.
Control Instruments Corporation
CSE Corporation
Day Star Corporation
Devco Engineering, Inc.
Digicolor
ERDCO Engineering Corporation
Foxboro Company
General Monitors, Inc.
Gow Mac Instrument Company
International Sensor Technology
Lumidor Safety Products
Macurco, Inc.
Matheson Gas Products, Inc.
Photovac, Inc.
Safety Supply Canada
Sensidyne, Inc.
Sentex Sensing Technology, Inc.
Sentrol Industrial, Inc.
Sierra Monitor Corporation
Thermo Environmental Instruments

Tracor Instruments

Infrared Analyzers and Accessories

Anacon
Astro Resources International Corporation
Jerome Instrument Corporation
Jones Safety Supply, Inc.
Laboratory Safety Supply Company
RS Landauer Jr. & Company
Lifecom Safety Service & Supply Company
Lumidor Safety Products
Mateson Chemical Corporation
Matheson Gas Products, Inc.
MC Products
MDA Scientific Inc.
Metrosonics, Inc.
Mine Safety Appliance Company
MSA
National Draeger Inc.
National Mines Service Company
Neotronics
Nuclear Associates
Nuclepore Corporation
Pendergast Safety Equipment
ProTech Safety Equipment, Inc.
Protective Equipment, Inc.
Raeco, Inc.
React Environmental Crisis Engineers

Safety Services, Inc.
Safety Supply Canada
Scott Specialty Gases
Sensidyne, Inc.
Sentrol Industrial, Inc.
Sierra Monitor Corporation
SKC, Inc.
SKC West, Inc.
Somatronix Research Corporation
Spectrex Corporation
Sperry Vision Corporation
Standard Marketing International, Inc.
Standard Safety Equipment Company
Supelco, Inc.
Syncor International Corporation
Syracuse Safety Services, Inc.
Tackaberry Company
Technical Associates
Teledyne Analytical Instruments
Texas Analytical Controls, Inc.
Thermo Analytical, Inc.
Thermonetics Corporation
3M Company, Occupational Health & Safety
Tierney Safety Products
Tornado Enterprises, Inc.
Vallen Safety Supply Company
Ward International
Wise El Santo Company
Xetex, Inc.

SOURCE: Occupational Health & Safety 1987/88 Purchasing Sourcebook, 2nd annual edition; Medical Publications, Inc., WACO, TX.

CHAPTER 17

MEDICAL SURVEILLANCE

I INTRODUCTION

It is difficult to assess with standard chemical monitoring procedures exposures that occur infrequently or briefly. It is also impossible with such procedures to identify the potentially adverse health effects of exposures occurring away from the workplace. Standard exposure monitoring cannot account for the effects of habits, such as smoking or drinking, nor of metabolic factors, such as variations in breathing rate, lung volume, or individual sensitivities. Medical surveillance and biological monitoring can help to fill in these gaps, making the health and safety program more comprehensive.

Biological monitoring involves monitoring chemicals, or metabolites of those chemicals, that have passed from the working environment into the biological or internal environment of the worker. Medical surveillance involves monitoring the health status of employees, with particular emphasis on the adverse health effects that may be caused by exposure to chemicals in the workplace.

Biological monitoring can be defined as "analysis of exhaled air, analysis of some biological fluid, such as urine, blood, tears, or perspiration; or analysis of some body component, such as hair or nails, to evaluate past exposure to a chemical" [1]. However, one can make distinctions between direct and indirect biological monitoring. Direct monitoring involves analysis of the chemical, or its immediate metabolites; indirect monitoring involves analysis of the effects that result from the action on some body system of the chemical or its metabolites. (The overlap in the definitions of indirect biological monitoring and medical monitoring can be attributed to a lack of information concerning the association between some observable secondary effects or physiological changes and the subsequent development of functional disorders.)

Medical surveillance also involves the analysis of body fluids or other components, but its fundamental purpose is to evaluate the pathological effects of past exposure. The objective is to reduce the level of morbidity and mortality attributed to the development of the subject disease in the population being monitored. The focus of medical monitoring is on establishing the probability of a disease being present, rather than on confirming the diagnosis of the disease. Thus, the tests that are conducted in a medical surveillance program tend to be simpler, less expensive, less invasive, and more comfortable than diagnostic test procedures.

II REGULATORY REQUIREMENTS FOR MEDICAL SURVEILLANCE

In regard to medical surveillance and biological monitoring, the regulations currently promulgated by the Occupational Safety and Health Administration (OSHA) are limited. They consist of the chemical-specific medical surveillance and biological monitoring requirements examples of which are summarized in Table 17–1.

OSHA has issued specific medical surveillance requirements for workers who wear respirators (see 29 CFR 1910.134 on respiratory protection), and within chemical-specific regulations (such as CFR 1910.1001, on asbestos). These regulations require periodic review of the medical status of respirator users to assure that they are physically able to perform their jobs while wearing the equipment. 29 CFR 1910.1001 also addresses issues that may arise if an employee is not able to wear a respirator, such as seniority status and pay rates, which cannot be jeopardized.

OSHA has also issued requirements for recordkeeping (29 CFR 1910.20) specifically distinguish between "exposure" records and "medical" records. These distinctions are particularly important for preserving the confidentially of medical records.

III ISSUES FOR PROGRAM IMPLEMENTATION

Before implementing a program of biological monitoring and medical surveillance, a laboratory must resolve a number of issues. Some are

TABLE 17-1   OSHA Medical Surveillance Requirements for Selected Chemicals

Chemical	Requirements
Asbestos	Pulmonary function tests, chest X-rays
Carcinogens (13)	Complete medical history (see 29 CFR 1910, 1003−1910, 1016), including genetic and environmental factors, review for immunologic status, medical treatments, pregnancy smoking history
Vinyl chloride	Complete physical exam, liver studies
Inorganic arsenic	Complete medical history and exam, chest X-ray, sputum cytology
Inorganic lead	Complete medical history and exam, detailed blood studies, including blood lead and zinc protoporphyrin levels
Coke oven emissions	Complete history, chest X-ray, pulmonary function tests, sputum cytology, urine cytology
Cotton dust	Complete medical history, standardized respiratory questionnaire, pulmonary function tests
1,2-Dibromo-3-chloropane	Complete medical and reproductive history, examination of genitourinary tract, serum specimen for radioimmunoassay
Acrylonitrile	Complete medical history and exam, detailed exam of peripheral and central nervous system, gastro-intestinal system, skin, and thyroid; chest x-ray; fecal occult blood screening for all workers over 40 years.

relatively simple, such as organization and administration. Others, such as some legal and ethical issues, or the selection of analytical techniques, are complex and may not be satisfactorily resolved with the current state of knowledge. Several general issues relate to the implementation of these programs.

The initial question a laboratory must answer is: Do we need a program? For example, if there are no exposures, or if the environmental monitoring is providing sufficient data for evaluating exposure, biological monitoring may not be necessary. On the other hand, there are many uses for biological monitoring data. For example, periodic biological monitoring may identify unsuspected exposures from intermittent sources or exposure routes — for example, absorption or ingestion — which traditional air monitoring techniques cannot detect. Since biological monitoring is a more direct measure of the dose received than is environmental monitoring, biological monitoring can confirm and document levels of exposure estimated from area or personal sampling. Medical personnel can use biological monitoring and medical surveillance data to confirm exposure and to guide the choice of therapeutics. They may also use such data to identify unusual sensitivities or metabolic abnormalities.

If a laboratory decides that monitoring or surveillance is worthwhile, it must decide who to include in the program, and it must establish frequency of surveillance. If cost were no concern, the laboratory could monitor all employees. However, a laboratory would likely choose to monitor or survey the employees whose exposures exceed certain thresholds (e.g., 50% of the TLV).

Next, a laboratory must answer these questions: What should be monitored? Should the chemical itself be monitored, or some metabolite, or some pathologic event? For example, if the laboratory monitors employees exposed to styrene, it will find almost all the styrene metabolized. The principal metabolite will be mandelic acid. In experimental animals, however, considerable interspecies differences occur, such as metabolism of styrene to hippuric acid rather than mandelic acid. Such differences complicate the development of biological monitoring methods. Human data are often lacking, and animal data may not lend themselves to the determination of a valid method.

Once the laboratory has determined what substance it wants to monitor, it must decide the kind of sample to take (e.g., blood, urine, breath, hair, nails) and specify the kind of analysis it wants (e.g., colorimetric, chromatographic, spectrophotometric). While blood is probably the best sampling medium to measure the exposure and function of vital organs, the sampling process is invasive. It requires special skill and procedures, both for sampling and processing, and may meet employee resistance. Other types of data, such as from analysis of urine or breath, are relatively easy to collect because they require no invasive techniques. However, since they are based on excretory samples, they are less accurate as a quantitative estimate of internal exposures.

The choice of analytical techniques varies from simple colorimetric "dipstick" tests to tests that require sophisticated equipment. Selection of a method depends on (1) the methods the laboratory has evaluated and considers to be acceptable, and (2) the availability of resources for in-house or commercial analysis.

After the analysis has been performed, the results must be interpreted. The laboratory must determine whether an exposure to the substance in question has occurred and, if so, whether the level of exposure falls within acceptable bounds. Unfortunately, the precision of biological monitoring and medical surveillance is frequently no better than ±20%. Interpretation of results can be difficult because individuals vary widely in their metabolic functions, dietary habits, and work and other exposures. Experienced occupational health professionals should be used to help design the program and to interpret the results. These results should be communicated to the employees and should be treated as confidential information.

IV CHEMICAL HYGIENE PLAN CONSIDERATIONS

The chemical hygiene plan must address medical surveillance and biological monitoring. Guidelines provided by the National Institutes of Health in "NIH Guidelines for the Laboratory Use of Chemical Carcinogens" [2] are useful in helping to develop the laboratory medical surveillance and biological monitoring program. Section III of the NIH guidelines, entitled Medical Surveillance, states that available practical procedures for detecting exposure to chemical carcinogens should be used for surveillance of employees who use these chemicals on a regular basis. Furthermore, the NIH guidelines state that the following information should be included in each employee's medical record:

1. reports of accidents that have resulted in:

    ingestion of chemical carcinogens,

    inhalation of chemical carcinogens, and

    any incidents that result in overt exposure of the employee to chemical carcinogens;

2. results of any surveillance procedures employed; and
3. results of environmental measurements of chemical carcinogens that may have been made within the employee's work environment.

The guidelines advise that each employee is to be informed of all results. In generalizing these guidelines for a laboratory handling hazardous materials other than carcinogens, the following checklist is suggested:

1. Medical examinations should be provided to personnel (including laboratory and maintenance workers) who will be working with or exposed to hazardous compounds or to animals, at the time they begin work, before they are exposed to potentially hazardous chemicals.

2. Follow-up medical examinations should be provided to these individuals at least every 12–18 months, and upon termination.

3. Laboratory (and maintenance) staff members who have to wear negative-pressure respirators to perform their work must have the approval of a physician for their use.

Where appropriate to the specific chemicals in use at the laboratory, biological monitoring programs may be implemented. Where such programs are used, they should be described in the hygiene plan. For example, monitoring erythrocyte cholinesterase activity is advisable where employees work with organophosphate pesticides. In such cases, the laboratory will typically monitor for exposure to pesticides such as carbaryl®, Diazinon®, malathion®, and parathion® by taking both pre- and postexposure samples. Postexposure levels of cholinesterase activity that are less than 70% of the preexposure baseline indicate exposure.

The chemical hygiene plan should also indicate that the medical surveillance and biological monitoring program will conform to applicable local, state, and federal regulations, and should summarize the requirements of those regulations applicable to the lab. (See Section II of this chapter.)

REFERENCES

1. R. S. Waritz, "Biological Indicators of Chemical Dosage and Burden," in *Patty's Industrial Hygiene and Toxicology*, Vol. III, L. J. Cralley and L. V. Cralley, Eds. New York: Wiley, 1979.

2. U.S. Department of Health and Human Services, National Institutes of Health, "NIH Guidelines for the Laboratory Use of Chemical Carcinogens," NIH Publication 81–2385, May 1981.

RESOURCES

Anderson, J. H., "Medical Aspects of Occupational Health in a Laboratory Setting," in *Laboratory Safety: Theory and Practice*, A. A. Fuscaldo, F. J. Erlick, and B. Hindman, Eds. New York: Academic Press, 1980.

Baselt, R. D., *Biological Monitoring Methods for Industrial Chemicals*. Davis, CA: Biomedical Publications, 1980.

Biotechnology, Inc., "OSHA Medical Surveillance Requirements and NIOSH Recommendations," prepared for the National Aeronautics and Space Administration, January 1980.

Haegele, L., "Selected Medical Problems Often Associated with laboratory Personnel," in *Laboratory Safety: Theory and Practice*, A. A. Fuscaldo, F. J. Erlick, and B. Hindman, Eds. New York: Academic Press, 1980.

Linch, A. L. *Biological Monitoring for Industrial Chemical Exposure Control*, Boca Raton, FL: CRC Press, 1974.

Messinger, H. B., R. Clappo, P. Nolan, and L. Stagner, "An Analysis of Medical Monitoring Data Required by OSHA Health Regulations," U.S. Department of Labor, Report ASPER/CON-78/0167/A, 1979.

Occupational Health: Recognizing and Preventing Work-related Disease, B. S. Levy, and D. H. Wegman (Eds.). Boston: Little Brown, 1983.

Piotrowski, J. K., "Exposure Tests for Organic Compounds in Industrial Toxicology," National Institute for Occupational Safety and Health (NIOSH), DHEW (NIOSH) Publication 77–144, 1977.

Rothstein, M. A., *Medical Screening of Workers*. Washington, DC: Bureau of National Affairs, 1984.

Rom, W. N., *Environmental and Occupational Medicine*. Boston: Little Brown, 1983.

U.S. Department of Health, Education, and Welfare, National Institute for Occupational Safety and Health, "Criteria for a Recommended Standard Occupational Exposure to Malathion," DHEW Publication (NIOSH) 76–205, 1976.

Young, G. S., "Laboratory Worker Medical Surveillance" in *Health and Safety for Toxicity Testing*, D. B. Walters, and C. S. Jameson, Eds. Boston: Butterworth, 1984.

CHAPTER 18

PROTECTIVE EQUIPMENT

I INTRODUCTION

Laboratory operations frequently involve a risk of clothing or skin contact with toxic materials. While engineering and administrative controls and good work practices reduce this risk, it is often necessary to augment those measures with personal protective equipment. However, unless workers select, use, and maintain that equipment properly, they may not receive the intended protection. Moreover, users may actually have a false sense of security and be at higher risk of injury or illness than if they had used no protective clothing at all.

If the first reason for using protective equipment in laboratories is the prevention of skin or clothing contact with hazardous chemicals, the second is contamination containment. Laboratories often adopt procedures requiring workers to don protective garments before entering a potentially contaminated area and remove them when exiting. This restricts any contamination that may have been picked up to one area.

The conditions under which protective equipment is used in a laboratory depend on the type of work being done, the chemical, biological, or physical agents being used, and the potential for exposure. This chapter addresses the use of protective equipment in a laboratory setting. It discusses the hazards against which protection is needed and the types of protective clothing available. The discussion in this chapter covers only hazards presented by skin and clothing contact; Chapter 19 discusses respiratory protection against inhalation hazards.

II THE USE OF PROTECTIVE CLOTHING IN LABORATORIES

A Protection of Face and Eyes

Many of the hazards for which protective clothing is required are chemical. The object of protective clothing is to form an impenetrable barrier between the user and the hazard. Unlike engineering controls, which enclose or otherwise affect the source of the hazard, one may think of protective clothing as enclosing the user against the hazard.

Direct contact of the skin or the mucous membranes with a chemical substance, or with clothing soaked with a chemical, will typically result in one of four adverse outcomes:

1. Inorganic acids and bases can cause primary irritation, ranging from minor reddening to ulceration and corrosion.
2. Organic solvents can cause defatting of the skin, with resultant cracking and dermatitis.
3. Some systemic toxins can be absorbed into the blood across the skin.
4. With some substances, a sensitization reaction may occur. In such cases, a much smaller exposure to the same substance at a later date can cause the reaction to occur again.

Dermatoses make up the majority of documented, occupationally related diseases that occur in the United States.

The most commonly used pieces of equipment for protection of the face and eyes are illustrated in Figure 18–1. Of these, the equipment most frequently used in laboratories is safety glasses. Safety glasses are available in many different configurations and styles, and can be fitted with prescription lenses for those who need them. Safety glasses can be equipped with side shields for greater protection.

Both glasses and goggles are available in tints and shades that permit their use for welding, burning, or for exposure to nonionizing radiation. In the case of lasers, the frequency of the radiation must be known, since the absorbing media are frequency-specific.

In environments in which a splashing liquid or rapidly moving airborne particulate may exist, clear plastic goggles that completely enclose the eyes will provide superior protection. In addition, goggles that can be

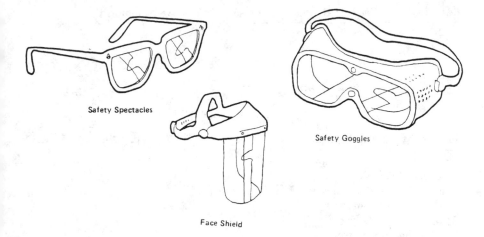

Safety Spectacles

Safety Goggles

Face Shield

FIG. 18–1: Personal protective equipment for protection of the face and eyes.

worn over nonsafety prescription glasses allow the use of "street" glasses in the lab.

A face shield provides protection for the nose and mouth areas of the face in addition to the immediate area of the eye (see Fig. 18–1). Clear shields are available for protection from splashing chemicals or rapidly moving particulate, such as one might encounter when using a grinding wheel. Finally, if total protection of the face, head, and neck is necessary, one may use a full hood with a clear visor.

Table 18–1 lists several leading suppliers and manufacturers of protective equipment for the eye and face.

B   Hard Hats and Hearing Protectors

While hard hats are not often seen in laboratories, in some applications such protection may be necessary. For example, many pilot plant operations require hard hats.

Hard hats are available in several models designed for different applications. Some hats are not appropriate for use near electrical wires, and others are only light-duty "bump caps," designed for use in low hazard environments.

Another form of personal protective equipment is exemplified by earplugs and earmuffs, which are available in a number of different types, sizes, shapes, and colors. As with all safety equipment, however, a qualified professional should select them and direct their use. Use of

TABLE 18−1 Selected Suppliers of Personal Protective Equipment

Supplier/Manufacturer	Air-Purifying Respirators	Air-Supplied Respirators	SCBA	Safety Glasses	Goggles	Face Shields	Hard Hats	Hearing Protection	Disposable Clothing	Reusable Clothing	Gloves
Abanda by Disposables, Inc.									√	√	
American Allsafe Company	√			√	√	√	√	√			
American Optical Company	√		√	√	√	√	√				
Bilsom, International					√		√	√			
H. L. Bouton Company				√	√	√					
Brahma Glove Company									√	√	√
E. D. Bullard Company	√					√	√				
Charkate									√	√	√
Daffin Disposables	√								√	√	√
Durafab									√	√	
EAR Division, Cabot Corporation								√			
Edmont									√	√	√
Glendale Protective Technologies	√			√	√	√	√	√			
Magid Glove & Safety Company	√		√	√	√	√	√	√	√	√	√
Mine Safety Appliance Company	√	√	√	√	√	√	√	√	√	√	√
North Safety Equipment	√	√	√	√	√	√	√				
Peltor, Inc.						√	√	√			
Racal Airstream, Inc.	√							√			
Rainfair, Inc.									√	√	√
Scott Aviation	√	√	√								
Survivair	√	√	√								
3M Company	√	√						√			
Titmus				√	√						
U.S. Safety, Cecso Service Company	√	√	√	√	√	√	√	√	√	√	√
Willson Safety Products	√	√		√	√	√	√	√			√

hearing protection is appropriate when one is working in areas with noise levels in excess of 85−90 dBA.

C Protective Clothing

Laboratory employees often wear protective clothing because there can be a high risk of spill or splash in laboratory operations. The protective clothing chosen must be appropriate to the hazard against which protection is required, and employees must use it according to manufacturer's directions (e.g., with respect to disposal, donning, cleaning).

Disposable laboratory coats, jumpsuits, and gloves often are found in laboratories. After a brief discussion of the other kinds of garment that may be applicable to the laboratory setting, the remainder of this section will focus on these items.

Protective clothing articles are made from a wide variety of natural and synthetic materials. Neoprene, nitrile, and natural rubbers, polyvinyl chloride, and polyethylene are a few examples of common protective clothing materials. Protective clothing designs vary, as well. One can purchase full-body suits, jackets, gloves, sleeves, aprons, and many other pieces of equipment. The selection of equipment appropriate to specific needs is discussed below.

D Disposable Laboratory Garments

Three basic types of fabric are in common use in industrial and laboratory disposable garments: most prevalent is Tyvek®, a spun-bonded polyolefin fabric made by du Pont. One drawback of Tyvek is that it is impermeable to air and moisture, and can be hot and uncomfortable. Manufacturers have developed a second class of breathable fabrics to address this problem. These fabrics, which include Gore-tex® and Kimberly-Clark's Kleenguard® fabric, are able to "breathe": that is, they will permit the passage of air and moisture and thus minimize retention of body heat. Such breathable garments may, however, also allow penetration of toxic agents. The laboratory should permit the use of such garments only after careful consideration of the protection requirements.

A final category of fabric is coated Tyvek, which features a layer of impermeable plastic. An example of the coating material used is Saranex®, which is manufactured by Dow.

All these fabrics are used to produce a wide variety of garments.

The choice of the correct garment requires a knowledge of the hazards involved and the protection required. Though one can assemble a complete protective ensemble by wearing several different garments at once, it is desirable to use as few different garments as possible to achieve the desired coverage. Whenever multiple garments are used, the possibility of chemicals entering the interior of the suit between layers is increased; one can minimize this risk by wearing single garments.

E Gloves

After the face and eyes, the hands are the most vulnerable part of the body, and are most likely to be affected by spills, cuts, accidental injections, or contact with temperature extremes. To mitigate each of these hazards, workers wear protective gloves. The selection of the right glove for the job requires a knowledge of the manual tasks to be performed, as well as an understanding of the physical and chemical risks to which the wearer may be exposed.

Tables 18–2 and 18–3 show how one glove supplier categorizes its products with respect to physical hazards and chemical hazards.

A more detailed discussion of some of the technical considerations that affect the selection and use of gloves and chemical protective clothing appears in Section III.

III SELECTION AND USE OF CHEMICAL PROTECTIVE CLOTHING

Since there is such a wide variety of chemical protective clothing available, and since the performance of various types against a particular hazard can differ greatly, the selection and use of chemical protective clothing must be carefully considered and administered. Consequently, chemical protective clothing must be selected by an individual who is knowledgeable in the types of uses of such gear. Among the issues that must be considered are permeability, functional compatibility, recordkeeping, and maintenance.

A Permeability

The extent to which a particular substance penetrates a glove material is called its permeability. In general, permeation of a substance through a glove or clothing material depends on at least four factors:

1. *Temperature.* The permeation rate increases with increasing temperature.

2. *Thickness.* The time required to permeate a material increases with the increasing thickness of a material.

3. *Solubility.* The permeability of a liquid is generally higher if the chemical is soluble in the protective garment material. However, the solubility cannot always be used as a predictor of permeability.

4. *Multicomponent liquids.* The rate at which a substance permeates can be accelerated in the presence of another, more rapidly permeating component.

Though qualitative permeation data are available from many manufacturers, relatively few have conducted any in-house quantitative exper-

TABLE 18−2 Chemical Protective Clothing Recommendations by Chemical Class (A Guide Not a Guarantee)

Chemical	Butyl Rubber	Chlorinated Polyethylene	Natural Rubber	Neoprene	Nitrile Rubber	Nitrile–PVC	Polyethylene	Polyurethane	Polyvinyl Acetate	Polyvinyl Chloride	Styrene–Butadiene Rubber	Viton
Acids Carboxylic												
Aliphatic and alicyclic, unsubstituted	R	r	**	RR	RR	R	NN	R	n	rr	R	r
Aldehydes												
Aliphatic	RR	rr	**	nn	nn	r		r	NN	rr	r	NN
Aromatic and heterocyclic	rr	**	nn	nn	nn	n		n	rr	N	n	**
Amides, carboxylic												
Aliphatic	**		**	nn	nn							nn
Amines Aliphatic and alicyclic												
Primary	**		**	rr	rr	r		**	rr	rr	**	**
Secondary	rr	nn	nn	**	nn	**		n	**	**	**	**
Tertiary	r	rr	rr	R	r	r			r	r		
Ester, carboxylic Aliphatic												
Acetates	**	r	NN	NN	NN	N	nn	**	rr	N	N	N
Higher Monobasic	r	nn			n					n		
Aromatic phthalates	**		nn	rr	rr				**			r
Ethers, aliphatic	nn	rr	nn	nn	rr	**		n	r	**	**	r

SOURCE: Guidelines for the Selection of Chemical Protective Clothing, American Conference of Governmental Industrial Hygienists, Cincinnati, OH.

[a] Legend: RR = recommended based on strong data.
rr = recommended based on data.
NN = not recommended based on strong data.

nn = not recommended based on data.
R = recommended based on judgment.
n = not recommended based on judgment.

203

TABLE 18–3 Physical Properties for Glove Use[a]

Property	Coated Work Gloves							Molded Handwear			
	Neoprene	Python Neoprene	Ripple Texture	Multi-purpose	Utility	Flexible Vinyl Plastic	Super-flexible Vinyl Plastic	Natural Latex Gloves	Latex Nitrile Gloves	Synthetic Bayprene® Rubber Gloves	Baytex® Gloves
Abrasion resistance	G	G	G	E	E	E	E	G	E	G	G
Cut resistance	E	E	E	NR	NR	NR	NR	E	E	E	E
Snag resistance	E	E	E	G	G	G	G	E	E	E	E
Heat resistance	G	G	E	F	F	F	F	E	G	G	E
Low temperature resistance	E	E	E	F	G	G	E	E	E	E	E
Flexibility	G	G	F	E	E	E	E	G	G	G	G
Dry grip	E	E	E	E	E	E	E	E	E	G	E
Wet grip	F	E	E	F	F	E	G	E	G	E	E

[a] G = good, E = excellent, F = fair, NR = not recommended.

ments. However, more permeability data are becoming available from both manufacturers and testing laboratories. The work by Schwope and Forsberg, cited in the Resources section at the end of this chapter, presents the most recent compilation of permeability information.

No material is 100% impermeable to anything, and no one material will form a satisfactory barrier against all substances. Thus, one must evaluate the performance of barrier materials against various chemicals on a substance-by-substance basis.

B Functional Compatibility

When prescribing various pieces of personal protective equipment for simultaneous use, it is useful to recall that if one kind of equipment is worn, the use of certain equipment may be unnecessary or impossible. For example, the use of a full-face respirator may obviate the need to use a face shield in some applications. Conversely, the use of prescription eye-glasses may preclude the use of a full-face respirator, since the temple bars would interfere with the sealing surface of the respirator. Finally, utilization of encapsulating suits, by definition, would require an alternative source of respirable air.

In selecting protective clothing, one must consider the hazards inherent in the work to be done. As an example, anyone who plans to work near an open flame or torch should wear flame-retardant garments.

C Recordkeeping and Administration

The incorrect use of safety equipment may place workers at higher risk of injury or illness. Those who assume themselves to be protected from a particular hazard when in fact they are not may take chances or fail to adopt a sufficiently cautious attitude. The result might be an injury that could have been avoided. For this reason, standard operating procedures should clearly describe the proper use of personal protective equipment. In addition, the laboratory should train its employees in its proper use and limitations, how to maintain it, and what to do in the event of a failure.

Furthermore, a laboratory must maintain complete records of employee training. A laboratory can use well-kept records to schedule annual refresher training, limit access to regulated areas, and facilitate issuance of protective clothing for a particular task.

The information presented here is just a general overview of the use of personal protective clothing. The reader may refer to other sources, particularly *Guidelines for the Selection of Protective Clothing* (prepared by Arthur D. Little, Inc., and cited in the Resources section under A. D. Schwope et al.) available from the American Conference of Governmental Industrial Hygienists (ACGIH), 6500 Glenway Avenue, Building D-5, Cincinnati, OH 45211; and "Chemical Protective Clothing Performance Index by K. Forsberg and L. H. Keith, available from John & Wiley Sons, New York, 10158 (1988).

IV EXAMPLES OF PROTECTIVE CLOTHING USE IN LABORATORIES

The protective equipment requirements for any laboratory must be tailored to the operations and hazards found in the lab. The following sections provide a few examples of protective equipment programs suitable for facilities of specific types.

For ancillary operations in toxicology laboratories (e.g., necropsy, tissue trimming, and histology operations), a laboratory employee should wear a single pair of disposable gloves, a laboratory coat, and safety glasses.

For work with carcinogens, mutagens, or teratogens, a worker should wear a disposable laboratory suit, safety glasses or goggles, gloves, disposable boots or shoe covers, sneakers or rubber boots, and a disposable head covering.

In chemical repositories handling toxic materials, personnel handling neat chemical should wear a disposable full-body suit over work clothing, as well as safety glasses, gloves, and shoe covers. Laboratory employees should not wear disposable overclothing out of the laboratory or repository where neat chemical is handled. They should remove work clothing upon exit from the laboratory.

For laboratory operations that do not involve the handling of neat hazardors chemical (e.g., chemical analyses, histology, tissue trimming, necropsy), workers should wear a single pair of disposable gloves, a laboratory coat, and safety glasses.

V CHEMICAL HYGIENE PLAN CONSIDERATIONS

The chemical hygiene plan should describe the protective equipment requirements of the laboratory. The work areas and/or tasks for which protective equipment is required should be specified. Equipment requirements for each area or task should be indicated. Responsibility for protective equipment selection, acquisition, cleaning, and disposal should be defined.

VI PROTECTIVE CLOTHING MANUFACTURERS AND DISTRIBUTORS

Some of the manufacturers and distributors of personal protective equipment were listed in Table 18—1. Laboratories may also refer to the Thomas Registry for a more comprehensive listing of manufacturers of a specific item.

RESOURCES

American National Standard for Men's Safety—Toe Footwear, ANSI 241, 1983.

American National Standard for Occupational and Educational Eye and Face Protection, ANSI 787.1, 1979. See also permeation guides published by protective clothing manufacturers (e.g., Seibe North, Edmont, Pioneer).

American National Standard for Protective Headwear for Industrial Workers, ANSI 789.1, 1981.

[All these are available from American National Standards Institute, 1430 Broadway, New York, New York 10018.]

Clayton, G. D., and E. E. Clayton (Eds.), *Patty's Industrial Hygiene and Toxicology*, 3rd rev. ed. New York: Wiley, 1978.

Forsberg, K., *Chemical Protective Clothing Performance Index with Gloves +*, Instant Reference Services, Austin, TX: 1989.

Forsberg, K., and L. H. Keith, *Chemical Protective Clothing Performance Index*, New York: Wiley, 1988.

Sansone, E. B., and L. A. Jonas, "The Effect of Exposure to Daylight and Dark Storage on Protective Clothing Material Permeability," *J. Am. Ind. Hygiene Assoc.*, 42, 841—843 (1981).

Schulle, H. E., "Personal Protective Devices," in *The Industrial Environment — Its Evaluation and Control*. Cincinnati, OH: National Institute for Occupational Safety and Health, 1973.

Schwope, A. D., et al., *Guidelines for the Selection of Chemical Protective Equipment*, 3rd ed. Cincinnati, OH: American Conference of Governmental Industrial Hygienists, 1987.

Williams, J. R., "Evaluation of Intact Gloves and Boots for Chemical Permeation," *Am. Ind. Hygiene Assoc. J.*, 42, 468—476 (1981).

CHAPTER 19

RESPIRATORS

I INTRODUCTION

Selection and use of respiratory protective devices raise a series of complex technical issues with which most laboratory personnel are unfamiliar. There is a wide variety of equipment available, and the regulations and recommendations affecting its use are continuing to evolve with changing technology. This chapter describes the requirements for a respirator program.

Respiratory protective devices can be divided into those that are air-purifying and those that are air-supplied. Alternatively, respirators can be classified as those that always maintain positive pressure within the face-piece and those that have negative pressure within the facepiece at some time in the breathing cycle.

An air-purifying respirator, as the name implies, protects the user by removing the contaminant(s) from inhaled air. The mechanism by which the contaminant is removed is substance-specific; that which effectively eliminates one contaminant will not necessarily do the same to another. An air-supplied respirator provides an external source of breathable air that is not affected by a hazardous environment.

A positive-pressure respirator maintains a constant greater-than-ambient facepiece pressure. In positive-pressure, air-purifying respirators, a fan provides a steady stream of purified air. In air-supplied positive-pressure

Table 19–1 Categorization of Respirators

Pressure	Air-Purifying	Air-Supplied
Negative	Disposable dust masks Half-mask or full-face cartridge	"Demand" air delivery mode
Positive	Powered air-purifying respirator (PAPR)	"Pressure-demand" air delivery mode Continuous-flow respirators

respirators, the delivery of air can be constant (continuous flow) or regulated, with air provided whenever the facepiece pressure drops below a certain, preset positive level (pressure demand).

On the other hand, in a negative-pressure respirator, the facepiece pressure drops below the ambient pressure whenever the user inhales. A problem arises if the face-to-facepiece seal is incomplete. The partial vacuum created by inhalation will allow unfiltered, contaminated air to enter the facepiece, fouling the wearer's air supply. Table 19–1 summarizes these two methods of categorizing respirators.

II KEY ELEMENTS OF A RESPIRATOR PROGRAM

A Written Respiratory Protection Program

The Occupational Safety and Health Administration (OSHA) requires any facility at which respirators are used to have a written respirator program that specifies and documents standardized procedures for the selection, assignment, use, and maintenance of respirators. The program, which should be available for ready reference by persons affected, should address:

- medical evaluation of respirator users,
- fit-testing procedures,
- training of employees,
- procedures for selecting respiratory protection,
- procedures for handling foreseeable emergencies,
- marking of air-purifying elements,
- air quality for supplied-air respirators,
- cleaning and maintenance, and
- procedures for disposal of contaminated air-purifying elements.

The written program serves as a handbook of the respirator program. The staff must revise it, as necessary, to reflect current practices, use situations, and changing regulations.

B NIOSH-Approved Respirators

The program must specify that laboratory staff may use only respirators certified by the National Institute for Occupational Safety and Health (NIOSH). Furthermore, the staff must use and maintain the respirators in a manner that is consistent with the manufacturer's instructions and recommendations.

A typical NIOSH certificate of approval is shown in Figure 19-1. For chemical cartridge, air-purifying respirators, the certificate indicates approval of the entire, assembled mask, including air-purifying elements. This procedure not only requires a separate certificate for each mask—cartridge combination, but it also precludes interchanging the parts of one manufacturer's respirator with those of another vendor.

For air-supplied respirators, a certificate indicates approval of entire ensembles of equipment from a single manufacturer. Interchanging parts, such as air lines or air cylinders, will void the certification, as it will for air-purifying equipment. For the special case of industrial fire brigades, OSHA does permit use of one make of air cylinder on another manufacturer's self-contained breathing apparatus [see 29 CFR 1910.156(f)(1)(iv)].

The NIOSH certificate explicitly lists the contaminants for which the certified equipment is effective and describes the conditions under which it may be used. The maximum use concentrations differ for three listed contaminants and this is a function of both the relative effectiveness of the equipment against the different substances and the relative toxicological characteristics of these materials. If there is uncertainty about the actual use concentration, one should err on the side of caution and opt for a more protective respirator.

The approval reproduced in Figure 19-1 applies to both a general class of contaminants (organic vapors, or dusts, fumes, and mists) and various specific contaminants (chlorine, hydrogen chloride, sulfur dioxide). In the case of approvals issued for a class of contaminants, one must exercise caution. The user should make sure that the actual use concentration does not exceed limitations based on the characteristics of the mask, as well as limitations based on the characteristics of the air-purifying element. This situation could arise when using the equipment approved in Figure 19-1 in an environment that contains an organic vapor with an OSHA-permissible exposure limit (PEL) of less than 100 ppm. Since the half-mask configuration purportedly reduces the exposure by a factor of

PERMISSIBLE RESPIRATOR CARTRIDGE FOR PESTICIDES

United States Department of Labor

MSHA

Mine Safety and Health Administration

MINE SAFETY AND HEALTH ADMINISTRATION

NATIONAL INSTITUTE FOR OCCUPATIONAL
SAFETY AND HEALTH

U S Department of Health Education and Welfare
Center for Disease Control

NIOSH

National Institute
for Occupational Safety and Health

APPROVAL NO. TC-23C-79

ISSUED TO Mine Safety Appliances Company, Pittsburgh, Pennsylvania, U.S.A.

LIMITATIONS

Approved for respiratory protection against pesticides. Do not wear in atmospheres containing less than 19.5 percent oxygen. Do not wear for protection against organic vapors with poor warning properties or those which generate high heats of reaction with sorbent materials in the cartridge. Maximum use concentrations will be lower than 1,000 parts per million organic vapors where that concentration produces atmospheres immediately dangerous to life or health. Not approved for fumigants.

CAUTION

In making renewals or repairs, parts identical with those furnished by the manufacturer under the pertinent approval shall be maintained.

Follow manufacturer's instructions for changing cartridges.

Refer to pesticide label for limitations on respirator use.

MSHA—NIOSH APPROVAL TC-23C-79 ISSUED TO Mine Safety Appliances Company, February 27, 1978

The approved half mask facepiece respirator assembly consists of the following MSA parts: 449703, 7-201-1 or 7-201-2 facepiece, and 448847 (TC-23C-79) cartridges. The approved half mask facepiece with belt mounted cartridge respirator assembly consists of the following MSA parts: 7-202 facepiece, breathing tube, and plenum assembly, and 448847 (TC-23C-79) cartridges. The approved full facepiece respirator assembly consists of the following MSA parts: 7-204 facepiece and 44847 (TC-23C-79) cartridges.

FIG. 19–1: Typical NIOSH certificate of approval.

10 (the protection factor), at a contaminant level of 1000 ppm the respirator will not prevent exposures exceeding the PEL. In the case of a toxic organic vapor with a PEL of 10 ppm, the maximum use specification is 100 ppm, not 1000 ppm. Persons familiar with respirator protection factors must be involved in equipment selection.

C Medical Evaluation of Respirator Users

The use of a respirator for protection can impose significant physiological and psychological stress on the wearer. Since such stress may place the individual at an elevated risk of injury or illness, a physician must evaluate the fitness of each person assigned to wear a respirator. Of course, the physician must be familiar with the conditions under which the respirator is to be worn.

Such an evaluation should include, but not be limited to:

- medical history, with special emphasis on cardiovascular or pulmonary disease;
- facial abnormalities that may interfere with a respirator seal;
- visual acuity;
- hearing ability;
- integrity of tympanic membranes;
- cardiovascular fitness;
- pulmonary function test; and
- other tests deemed appropriate by the physician (e.g., endocrine system, psychological status, neurological health, exercise stress tests).

After the evaluation, the physician should provide a written statement of the results of the exam, including whether the person is medically qualified to use a respirator and, if so, under what limitations.

D Fit-Testing Procedures

The laboratory must fit-test all employees who are required to wear negative-pressure respirators to determine which mask best conforms to their facial features. A fit test is a rigorous protocol in which the tester challenges the face-to-facepiece seal with a chemical agent. Detection of the chemical agent inside the facepiece indicates the presence of a leak. Appendix A to this chapter presents the recommended protocol for the isoamyl acetate (IAA) qualitative fit test.

To be valid, the isoamyl acetate qualitative fit test requires a voluntary, truthful answer with regard to detection of the test agent inside the facepiece. If the tester suspects that a truthful response was not given, or if the test subject is insensitive to isoamyl acetate, the tester may revert to the irritant smoke qualitative fit test. One should administer this test with caution, however, because the response of the test subject to the irritant smoke inside the facepiece is both involuntary and unpleasant. Also, it may involve some coughing and gagging. Appendix B presents a protocol for administration of this test.

A laboratory should fit test employees before initial assignment to any job that may require the use of a respirator, and at least annually thereafter. More frequent fit-testing is necessary for new mask configurations, or if an employee's facial contours change radically from weight loss, injury, or illness.

Another factor that may affect the fit and protection afforded by a respirator is the presence of facial hair between the mask and the surface of the face. Facial hair will permit the passage of unpurified air into the interior of the facepiece, and the wearer will assuredly inhale it. For this reason, laboratory staff with facial hair cannot rely on negative-pressure respirators utilizing tight-fitting half- or full-facepieces. Alternatives appropriate for use by persons with facial hair include loose-fitting hoods, like a 3M Whitecap, or a shroud that covers the head and upper torso. Also, a positive-pressure (pressure-demand), air-supplied respirator with a tightly fitting facepiece is a viable alternative.

E Training

The quality and the quantity of training provided to respirator users are critical in determining the level of protection afforded in a given use situation. At a minimum, the laboratory should offer appropriate training on initial assignment, and at least annually thereafter, or whenever the potential for exposure changes.

Training should provide information with regard to:

- functional components,
- preuse inspection,
- air-purifying element selection,
- donning instructions,
- functional checks for positive and negative pressure,
- limitations,
- typical use situations,

- emergency instructions,
- care and maintenance, and
- storage locations.

A laboratory can organize its training into three sections: preuse instruction, instructions for normal and emergency use, and maintenance and care instructions.

Preuse instructions will ensure the user that the respirator is working properly, that all necessary functional components are present, and that he or she has donned the respirator properly. The user must understand the function and assembly of a respirator and be able to perform a preuse inspection.

In preuse instruction, the training leader shows how to don the respirator, including proper orientation of the facepiece, proper strap tension, and correct hose and valve setup (if applicable). The user then learns how to adjust wear comfort based on facepiece pressure and strap tension and, finally, how to perform the positive- and negative-pressure functional checks that are used to test for any leakage in the face-to-facepiece seal.

The purpose of the positive- and negative-pressure functional checks is to ensure proper mask seal and to verify that all the necessary parts are present and operating properly. The positive-pressure test consists of sealing the exhalation valve cover with one's hand and exhaling gently. If the respirator lifts up slightly off the face without leaking, the respirator passes the test.

The negative-pressure test is conducted by covering the air inlet(s) and gently inhaling. The respirator passes if the mask collapses slightly against the face without leaking.

Neither pressure test qualifies as a qualitative fit test, and neither can be used as a basis for respirator assignments. However, both are important daily user checks.

Second, training should be specific to the situation in which the trainee will wear the respirator. The user must become familiar with and conversant in standard operating procedures. These include entry and exit from the contaminated area, donning and doffing the respirator, and selection and replacement of the air-purifying element (if applicable).

To the extent that respirators are used in emergencies, including entry into or escape from contaminated environments, training must be specific to special emergency procedures. This training should emphasize the limitations of air-purifying equipment in emergency situations and should provide detailed instructions on any supplied-air equipment or self-contained breathing apparatus (SCBA) that may be used.

F   Selection of Respiratory Protection

In selecting the correct respirator for a task, one should consider the following:

- type of hazardous environment,
- chemical—physical characteristics of the hazard,
- acute and chronic health effects of exposure,
- concentration of contaminant and ambient oxygen,
- warning properties,
- ease of escape from contaminated area,
- necessity for skin and eye protection,
- length of time for which respirator must be worn, and
- work activities and characteristics.

The qualified professional who selects the respirator for a particular hazardous environment should be familiar not only with the hazards involved, but also with the capabilities of the respirator and the needs of the user. The laboratory should review the criteria for such choices regularly.

NIOSH has compiled many of the decisions that go into the selection process into a document, entitled the "NIOSH Respirator Decision Logic." Use of the decision logic provides a rigorous framework within which one can progressively exclude inappropriate respirators until only correct respirators remain under consideration.

OSHA has published a second decision logic as part of its *Industrial Hygiene Technical Manual*. It is included as Figure 19−2.

G   Maintenance and Care

When respirators are assigned to persons for their exclusive use, the user is responsible for routine maintenance and care and for functional checks, including positive- and negative-pressure tests before each use. Respirator cleaning and disinfecting and proper storage after each use are also responsibilities of the individual.

Finally, respirators that have been designated for general or emergency use must be inspected at least monthly, and cleaned and disinfected after each use.

Respirator manufacturers specify correct cleaning procedures. Typically, for cleaning, they recommend that the user break down the respirator into its component parts, wash them thoroughly with a mild detergent,

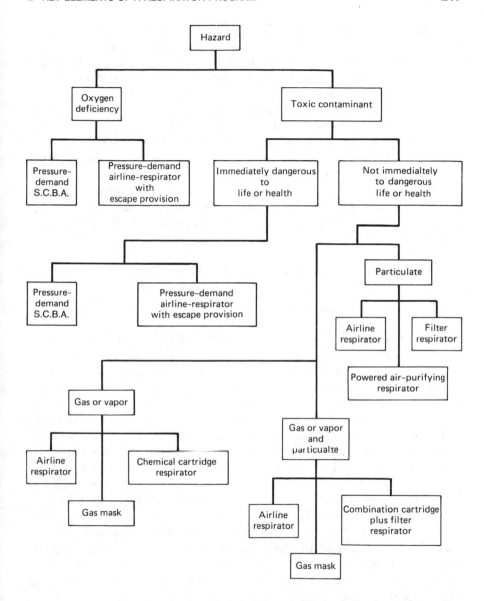

SOURCE: OSHA Industrial Hygiene Technical Manual. Chapter 5: Respiratory Protection.

FIG. 19−2:   Respiratory selection for routine use of respirators.

rinse thoroughly, and allow the equipment to dry in the air. They also recommend use of a drying cabinet to minimize drying time.

Upon reassembly, the user should carefully inspect the respirator and replace any worn or missing parts, paying particular attention to the inhalation and exhalation valves, the condition of the facepiece, the elasticity of the straps, and the presence of any rusted metal parts. Users should also check the date on the air-purifying element and, if necessary, replace the old elements with fresh ones. Finally, they should sanitize the respirator with a nonalcoholic disinfectant pad, seal it in a clean plastic bag, and store it in a cool, dry place until needed.

The replacement of old or depleted air-purifying elements is part of the normal care and maintenance of a respirator. In cases of cartridge/canister units that may be safely reused, depletion of the air-purifying capacity is indicated by:

- penetration of the substance through the cartridge (breakthrough),
- increase in the resistance to breathing (overloading), and/or
- a change in the end-of-service-life indicator (ESLI).

An end-of-service life indicator is a visible signal that the air-purifying capability of a cartridge or canister is, or is not, viable. An ESLI is particularly important for substances with poor warning properties: carbon monoxide, for example, is not detectable via smell or taste in the event of a cartridge/canister breakthrough. Upon evidence of any of the signs just listed, the user should replace the cartridge or canister without delay.

However, the user should anticipate the need to change the cartridge or canister, regularly replacing old or used air-purifying elements. The frequency of replacement can vary from weekly, for high use respirators, to annually for respirators used only in emergencies. In any case, one should place the service date on the side of the cartridge or canister.

Special disposal procedures may be necessary when the air-purifying elements, by virtue of the contaminants in question, become a possible source of exposure after use. Such may be the case with cartridges used to clean up a hazardous material spill. In such instances, it may be necessary to treat the contaminated cartridges or canisters as hazardous waste and dispose of them accordingly.

H   Air Quality for Air-Supplied Respirators

The immediate source of air for air-supplied respirators, typically, is a mechanical compressor or a tank of compressed breathing air. In either case, a laboratory must take steps to ensure that unwanted contaminants are not present in the air.

Where a mechanical compressor supplies the breathing air, one must first remove the oil, water, and carbon monoxide from the air. The carbon monoxide treatment consists of a catalytic conversion of carbon monoxide to carbon dioxide. Since moisture can render a catalyst ineffective, a safe procedure calls for installing a carbon monoxide alarm downstream of the catalyst to alert users of contaminated air.

A laboratory should ensure that the intake of the compressor is not located near any potential contaminant sources. In such a case, if the compressor introduced a toxic substance into the breathing air, the purification system would be able neither to filter it out nor detect it.

Since an appreciable pressure drop exists across the air purification system, inlet air must be induced at a sufficiently high pressure to ensure that the downstream pressure meets the manufacturer's specifications for the respirator.

Even when using a self-contained breathing apparatus or breathing air cylinder, air quality is a major concern. Users of these air cylinders should make sure that the supplier is aware of the need for air purification. The laboratory should request certificates of analysis with every shipment of cylinders, and only air that meets "breathing air" quality standards should be used.

Finally, as a precaution against using nonbreathing air cylinders, the hose fittings for air-supplied respirators must be absolutely incompatible with hose fittings for other gases.

III   RECORDKEEPING AND DOCUMENTATION

Rigorous recordkeeping, which is necessary to document compliance with the written program, is also an invaluable management tool. The program administrator should maintain records of the following:

- training
- medical evaluations, including a physician's written statement of approval or nonapproval to wear respirators (if limitations on use apply, they should be stated);
- fit-testing;
- maintenance;
- written respiratory protection program and pertinent amendments; and

- standard operating procedures for specific routine and foreseeable use situations, and pertinent amendments.

IV CHEMICAL HYGIENE PLAN CONSIDERATIONS

The chemical hygiene plan should include (either directly or by reference) the laboratory's respirator program. The program should include:

- Specific indication of who is responsible for each program element.
- Requirement that only NIOSH-approved respirators selected by a qualified individual for protection against the hazards of the area be used.
- Training on respirator use for those who may be called on to wear respirators.
- Fit-testing of respirator users.
- Medical approval for respirator use for those who may be called on to wear respirators.
- Procedures for issuance, inspection, maintenance, and cleaning of equipment.
- When self-contained breathing apparatus is to be available for emergency use, procedures for inspection and maintenance of this equipment.
- Listing of when respirators are to be used. Laboratory staff members should use respirators:

  as a secondary exposure control strategy in conjunction with engineering controls;

  as a temporary exposure control strategy when engineering controls are being installed, or are otherwise unavailable;

  as a primary means of controlling exposure when engineering controls are not feasible; and

  during emergencies.

APPENDIX A
ISOAMYL ACETATE (BANANA OIL) QUALITATIVE FIT-TEST PROTOCOL

  I. *Respirator Selection Procedure*
  Purpose: To select the most comfortable respirator from a group of

possible choices; to evaluate comfort over an extended period of time; and to instruct test subjects in the positive-pressure and negative-pressure functional checks.

A. Based on facial size and shape, the subject shall be allowed to select the most comfortable respirator from a selection, including respirators of various sizes from different manufacturers.

B. The test subject should understand the purpose and function of the respirator, and be shown how to don it correctly, how to tighten and position the straps, and how to assess the comfort of the fit.

This summary instruction does not satisfy the training requirements outlined in Section II, E.

C. Once the most comfortable respirator has been selected and donned, the test subject shall wear it for an extended period of time (at least 5 minutes) to assess the comfort over prolonged use.

D. The test subject shall conduct the conventional positive- and negative-pressure tests. Prior to conducting these tests, the subject shall "seat" the respirator by shaking his head back and forth and up and down while breathing deeply.

E. The purpose of the positive- and negative-pressure functional checks is to ensure that the mask is sealed properly and that all the necessary parts are present and operating properly. The positive-pressure test is conducted by sealing the exhalation valve cover with one's hand and exhaling gently. If the respirator lifts up slightly off the face without leaking, then the respirator passes the test.

The negative pressure test is conducted by covering the air inlet(s) and gently inhaling. The respirator passes if the mask collapses slightly against the face without leaking.

Neither the positive- nor the negative-pressure test qualifies as a qualitative fit test, and thus neither can be used as a basis on which respirator assignments can be made.

F. If the respirator fails either the positive- or the negative-pressure test, or becomes uncomfortable, another respirator should be selected and steps C, D, and E performed again.

II. *Odor Threshold Screening Procedure*

Purpose: To familiarize the test subject with the test procedure and the odor of isoamyl acetate; and to identify subjects who are insufficiently sensitive to the odor of isoamyl acetate to participate.

A. The *stock solution* is made by adding 1 cc of pure isoamyl acetate to 800 cc of odor-free water in a 1-quart glass jar with metal lid. Shake vigorously for 30 seconds. The stock solution has a shelf-life of one week.

B. The *odor test solution* is prepared by adding 0.4 cc of the stock solution to 500 cc of odor-free water in a 1-quart glass jar with metal lid. Shake vigorously for 30 seconds. The odor test solution must be prepared fresh every day it is to be used.

C. The *blank solution* is prepared by placing 500 cc of odor-free water in a 1-quart glass jar identical to that containing the odor test solution.

D. The jars containing the odor test and the blank solutions shall be labeled so that only the test administrator can identify them. The labels shall be changed periodically to preserve the integrity of the test.

E. Steps A−D shall be performed in a room separate from that used for Step F.

F. The subject shall be asked to shake each bottle, to remove the lids one at a time, and to identify the bottle which contains the odor test solution. Only those subjects who correctly identify the jar containing the odor test solution may participate in the actual fit test.

III. *Fit Testing*

A. The fit-test chamber shall be similar to a clear 55-gallon drum liner inverted over a 2-foot diameter from suspended so that the subject's head is 6−8 inches below the top of the chamber.

B. The respirator that was selected as being most comfortable shall be equipped with organic vapor cartridges for the test. The test subject should perform the positive- and negative-pressure functional checks.

C. The room in which the fit-test chamber is located shall be well-ventilated to prevent general room contamination, which can lead to odor fatigue.

D. Upon entering the fit-test chamber, the test subject shall be given a 6″ × 5″ piece of single-ply paper towel, folded once, and wetted with 0.75 cc of pure isoamyl acetate. The test subject shall hang the paper towel from a hook at the top of the fit-test chamber.

E. After allowing 2 minutes for the isoamyl acetate concentration to build up inside the chamber, the test subject should perform the following exercises:

1. Breathe normally.
2. Breathe deeply, with regular breaths.
3. Turn head from one side to the other, breathing on each side.
4. Nod head up and down, breathing while head is up and also while head is down.
5. Have the test subject read the following "Rainbow Passage," a copy of which should be posted on the inside of the fit-test chamber:

"When sunlight strikes raindrops in the air, they act like a prism and form a rainbow. The rainbow is a division of white light into many beautiful colors. These take the shape of a long round arch, with its path high above, and its two ends apparently beyond the horizon. There is, according to legend, a pot of gold at one end. People look, but no one ever finds it. When a man looks for something beyond his reach, his friends say he is looking for the pot of gold at the end of the rainbow."

6. If the subject has difficulty reading the rainbow passage, have him recite the alphabet and count slowly for 30 seconds.
7. Breathe normally.
F. If the odor of isoamyl acetate is detected at any time during the procedure, the test subject should immediately leave the chamber to prevent olfactory fatigue.
G. If the subject cannot pass the test using one of the half-mask respirators available, the test shall be repeated using full-face masks.

SOURCE: Draft proposed OSHA 29 CFR 1910.134, July 20, 1985.

APPENDIX B
IRRITANT SMOKE QUALITATIVE FIT-TEST PROTOCOL

I. *Respirator Selection Procedure*
   A. Respirators shall be selected as described [in Section I of Appendix A], except that each respirator shall be equipped with combination acid gas−organic vapor/high-efficiency particulate cartridges.
II. *Fit-Testing*
   A. The test subject shall be allowed to smell weak concentrations of isoamyl acetate and irritant smoke to familiarize him with the characteristic odor of each.
   B. The test subject shall properly don the respirator selected as above, and wear it for at least 10 minutes before starting the fit test.
   C. The test conductor shall review this protocol with the test subject before testing.
   D. The test subject shall perform the conventional positive-pressure and negative-pressure fit checks. Failure of either check shall be cause to select an alternate respirator.
   E. A simplified isoamyl acetate-based fit test shall be performed using ampules of IAA. Pass the ampule around the perimeter of the respirator at the junction of the facepiece and face. If leakage is detected, readjust the respirator. If leakage persists the respirator is rejected.
   F. If no odor of IAA is detected, the irritant smoke test shall be administered.
   G. Break both ends of a ventilation smoke tube that contains stannic oxychloride, such as the MSA [Mine Safety Appliances Co.] part No. 5645, or equivalent. Attach a short length of tubing to one end of the smoke tube. Attach the other end of the smoke tube to a low-pressure air pump set to deliver 200 milliliters per minute.
   H. Advise the test subject that the smoke can be irritating to the eyes and instruct him to keep his eyes closed while the test is performed.
   I. The test conductor shall direct the stream of irritant smoke from the tube towards the face seal area of the test subject. He shall begin at

least 12 inches from the facepiece and gradually move to within 1 in, moving around the whole perimeter of the mask.

J. The following exercises shall be performed while the respirator seal is being challenged by the smoke. Each shall be performed for one minute.

1. Normal breathing.
2. Deep breathing. Be certain breaths are deep and regular.
3. Turning head from side to side. Be certain movement is complete. Alert the test subject not to bump the respirator on his/her shoulders. Have test subject inhale when his/her head is at either side.
4. Nodding head up and down. Be certain motions are complete. Alert the test subject not to bump the respirator on his/her chest. Have test subject inhale at the extremes of the range of motion.
5. Talking slowly and distincly, have test subject count backward from 20.
6. Normal breathing.

K. If the irritant smoke produces an involuntary response, such as coughing, the test conductor shall consider that the tested respirator does not fit.

L. Each test subject passing the smoke test without evidence of a response shall be given a sensitivity check of the smoke from the same tube to determine whether he reacts to the smoke. Failure to evoke a response shall void the fit test.

M. Steps B4, B9, B10 of this protocol shall be performed in a location with exhaust ventilation sufficient to prevent general contamination of the testing area by the test agents (IAA, irritant smoke).

N. Respirators successfully tested by the protocol may be used in contaminated atmosphere up to 10 times the established exposure limit.

SOURCE: Draft proposed OSHA 29 CFR 1910.134, July 20, 1985.

CHAPTER 20

CHEMICAL INCOMPATIBILITY

I INTRODUCTION

The risks associated with chemical incompatibilities must be managed in any laboratory activity in which chemicals are handled or used. The mixing of incompatible chemicals can result in sudden, violent, and unforeseen hazards and may cause significant personal injury and property damage.

This chapter briefly addresses the types of risk that attend the mixing of incompatible chemicals and provides some resources for the investigator trying to avoid such hazards.

II DANGERS OF INCOMPATIBLE CHEMICALS

In general, chemicals react to form compounds or other chemicals, with an attendant consumption or generation of energy. The dangers that are inherent in chemical incompatibility occur when the end products or by-products themselves are toxic or hazardous, or when the generation of energy is at a magnitude to be destructive.

Dangerous by-products or reaction products include solids, liquids, and gases, with the latter being of greatest concern. Chemical reactions can produce gases and vapors that are harmful by virtue of their toxicity or their flammability, or both. Even the reactions that generate a substantial amount of an innocuous gas or vapor warrant concern, since they can displace the available oygen in an enclosed area and create an oxygen-deficient environment.

The formation of liquid or solid hazardous reaction products can also present risks. Reagents that are thought to be pure can be used for applications that might be inappropriate for the liquid or solid reaction product.

The generation of energy is often accompanied by the generation of toxic reactions products. A fire will produce not only light and heat, but also the toxic products of combustion. Whenever the generation of light, heat, or pressure occurs in excessive magnitude, or with excessive speed, an explosion or fire can result, and the effect can be catastrophic.

III AVOIDING INCOMPATIBLE CHEMICAL REACTIONS

The contact and subsequent reaction of incompatible chemicals can occur via one of two paths. Either the two can mix as a result of an accident, or their intentional combination in the laboratory can go awry.

Avoiding the accidental mixture of chemicals is accomplished in the same manner as the avoidance of spills and leaks (discussed elsewhere in this manual). The special precautions outlined in Chapter 8 for storage of hazardous materials are further supported by the need to avoid the mixture of incompatible chemicals. The segregation of incompatible materials in storage areas is particularly important. If the same accident or event (e.g., knocking over a shelf or an earthquake) could cause containers of incompatible materials to break and their contents to mix, one of the problems outlined above could result.

The other avenue by which adverse chemical reactions may occur involves the intentional mixing of chemicals in the laboratory. This activity must always be preceded by careful analysis of the materials being used, and the possibility of an unwanted or unforeseen reaction. Reference must be made to the material safety data sheet (MSDS), if one is available, or to the incompatibility information contained in health and safety reference sources. In addition, charts like those presented in Figures 20−1 and 20−2 and Table 20−1 should be consulted.

If preexperimental research indicates the potential for the reaction of incompatible material, the laboratory must prepare and implement special operating procedures to manage the risk.

The following pages list the chemicals by chemical name and by reactivity groups. Obtain the group for the chemical and then read chart, first from left to right, then down.

Chemicals Not On Chart

Carbon Bisulfide forms an unsafe combination with reactivity groups 1, 4, 19, 20, and epichlorohydrin.

Epichlorohydrin forms an unsafe combination with reactivity groups 1, 2, 3, 4, 14, 15, 19, 20, 22, 23, 24, and carbon bisulfide.

Motor Fuel antiknock compounds form unsafe combinations with reactivity groups 1, 4, 5, 6, 7, 15, 19, and 20.

	1	2	3	4	5	6	7	8	9	10	11	12	13	14	15	16	17	18	19	20	21	22	23	24
1 Inorganic Acids																								
2 Organic Acids	X																							
3 Caustics	X	X																						
4 Amines & Alkanolamines	X	X	X																					
5 Halogenated Compounds	X			X																				
6 Alcohols, Glycols & Glycol Ethers	X																							
7 Aldehydes	X	X	X	X		X																		
8 Ketones	X		X	X																				
9 Saturated Hydrocarbons																								
10 Aromatic Hydrocarbons	X																							
11 Olefins	X				X																			
12 Petroleum Oils																								
13 Esters	X		X	X																				
14 Monomers & Polymerizable Esters	X	X	X	X	X	X																		
15 Phenols			X	X																				
16 Alkylene Oxides	X	X	X	X		X	X	X					X	X	X									
17 Cyanohydrins	X		X	X	X	X	X																	
18 Nitriles	X		X	X																				
19 Ammonia	X	X																						
20 Halogens	X		X			X	X	X	X	X	X		X	X	X	X	X	X	X					
21 Ethers	X																							
22 Phosphorus, Elemental	X	X	X																	X	X			
23 Sulfur, Molten																				X		X		
24 Acid Anhydrides	X		X	X		X	X	X	X	X	X	X	X	X	X	X	X	X	X	X				

Source: CHRIS Hazardous Chemical Data.

FIG. 20–1: Chemical compatibility chart.

FIG. 20—1 (Continued)

Reactivity Groups.

Group 1: Inorganic Acids

Chlorosulfonic acid
Hydrochloric acid (aqueous)
Hydrofluoric acid (aqueous)
Hydrogen chloride (anhydrous)
Hydrogen fluoride (anhydrous)
Nitric acid
Oleum
Phosphoric acid
Sulfuric acid

Group 2: Organic Acids

Acetic acid
Butyric acid (*n*-)
Formic acid
Propionic acid
Rosin oil
Tall oil

Group 3: Caustics

Caustic potash solution
Caustic soda solution

Group 4: Amines and Alkanolamines

Aminoethylethanolamine
Aniline
Diethanolamine
Diethylamine
Diethylenetriamine
Diisopropanolamine
Dimethylamine
Ethylenediamine
Hexamethylenediamine
Hexamethylenetetramine
2-Methyl-5-ethylpyridine
Monoethanolamine
Monoisopropanolamine
Morpholine
Pyridine
Triethanolamine
Trietylamine
Triethylenetetramine
Trimethylamine

Group 5: Halogenated Compounds

Allyl chloride
Carbon tetrachloride
Chlorobenzene
Chloroform
Chlorohydrins, crude
Dichlorobenzene (*o*-)
Dichlorobenzene (*p*-)
Dichlorodifluoromethane
Dichloroethyl ether
Dichloropropane
Dichloropropene
Ethyl chloride
Ethylene dibromide
Ethylene dichloride
Methyl bromide
Methyl chloride
Methylene chloride
Monochlorodifluoromethane
Perchloroethylene
Propylene dichloride
1,2,4-Trichlorobenzene
1,1,1-Trichloroethane
Trichloroethylene
Trichlorofluoromethane

Group 6: Alcohols, Glycols and Glycol Ethers

Allyl alcohol
Amyl alcohol

1,4-Butanediol
Butyl alcohols (iso, *n*, sec, tert)
Butylene glycol
Corn syrup
Cyclohexyl alcohol
Decyl alcohols (*n*, iso)
Dextrose solution
Diacetone alcohol
Diethylene glycol
Diethylene glycol dimethyl ether
Diethylene glycol monobutyl ether
Diethylene glycol monoethyl ether
Diethylene glycol monomethyl ether
Diisobutyl carbinol
Dipropylene glycol
Dodecanol
Ethoxylated dodecanol
Ethoxylated pentadecanol
Ethoxylated tetradecanol
Ethoxylated tridecanol
Ethoxytriglycol
Ethyl alcohol
Ethyl butanol
2-Ethylbutyl alcohol
2-Ethylhexyl alcohol
Ethylene glycol
Ethylene glycol monobutyl ether
Ethylene glycol monoethyl ether
Ethylene glycol monomethyl ether
Furfuryl alcohol*
Glycerine
Heptanol
Hexanol
Hexylene glycol
Isoamyl alcohol
Isooctyl alcohol
Methoxytriglycol*
Methyl alcohol
Methylamyl alcohol
Molasses, all

Nonanol
Octanol
Pentadecanol
Polypropylene glycol methyl ether
Propyl alcohols (*n*, iso)
Propylene glycol
Sorbitol
Tetradecanol
Tetraethylene glycol
Tridecyl alcohol
Triethylene glycol
Undecanol

Group 7: Aldehydes

Acetaldehyde
Acrolein (inhibited)
Butyraldehyde (*n*, iso)
Crotonaldehyde
Decaldehyde (*n*, iso)
2-Ethyl-3-propylacrolein
Formaldehyde solution
Furfural
Hexamethylenetetramine
Isooctyl aldehyde
Methyl butyraldehyde
Methyl formal*
Paraformaldehyde
Propionaldehyde
Valeraldehyde

Group 8: Ketones

Acetone
Acetophenone
Camphor oil
Cyclohexanone
Diisobutyl ketone
Isophorone*
Mesityl oxide*

* Not presently included in CHRIS system.

FIG. 20–1 (Continued)

Methyl ethyl ketone
Methyl isobutyl ketone

Group 9: Saturated Hydrocarbons

Butane
Cyclohexane
Ethane
Heptane (n-)
Hexane (n, iso)
Isobutane
Liquefied natural gas
Liquefied petroleum gas
Methane
Nonane
n-Paraffins
Paraffin wax
Pentane (n, iso)
Petrolatum
Petroleum ether
Petroleum naphtha
Polybutene
Propane
Propylene butylene polymer

Group 10: Aromatic Hydrocarbons

Benzene
Cumene
p-Cymene*
Coal tar oil
Diethylbenzene
Dodecyl benzene*
Dowtherm
Ethyl benzene
Naphtha, coal tar
Naphthalene (includes molten)
Tetrahydronaphthalene
Toluene

Triethyl benzene
Xylene (m-, o-, p-)

Group 11: Olefins

Butylene
1-Decene
Dicyclopentadiene
Diisobutylene
Dipentene*
dodecene
1-Dodecene
Ethylene
Liquefied petroleum gas
1-Heptene
1-Hexene
Isobutylene
Nonene
1-Octene
1-Pentene
Polybutene
Propylene
Propylene butylene polymer
Propylene tetramer (dodecene)
1-Tetradecene
1-Tridecene
Turpentine
1-Undecene

Group 12: Petroleum Oils

Asphalt
Gasolines
   Casingead
   Automotive
   Aviation
Jet fuels

* Not presently included in CHRIS system.

JP-1 (kerosene)
JP-3
JP-4
JP-5 (kerosene, heavy)
Kerosene
Mineral spirits
Naphtha (non aromatic)
Naphtha
  solvent
  Stoddard solvent
  VM & P
Oils
  Absorption oil
  Clarified oil
  Crude oil
  Diesel oil
  Fuel oils
    No. 1 (kerosene)
    No. 1-D
    No. 2
    No. 2-D
    No. 4
    No. 5
    No. 6
  Lubricating oil
  Mineral oil
  Mineral seal oil
  Motor oil
  Penetrating oil
  Range oil
  Road oil
  Spindle oil
  Spray oil
  Transformer oil
  Turbine oil*

Group 13: Esters

Amyl acetate
Amyl tallate

Butyl acetates (n, iso, sec)
Butyl benzyl phthalate
Castor oil
Cottonseed oil
Croton oil*
Dibutyl phthalate
Diethyl carbonate
Dimethyl sulfate
Dioctyl adipate
Dioctyl phthalate
Epoxidized vegetable oils
Ethyl acetate
Ethyl diacetate
Ethylene glycol monoethyl ether acetate
ethylhexyl tallate
fish oil
Glycol diacetate
Methyl acetate
Methyl amyl acetate
Neatsfoot oil
Olive oil
Peanut oil
Propyl acetates (n, iso)
Resin oil
Soya bean oil
Sperm oil
Tallow
Tanner's oil
Vegetable oil
Wax, carnauba

Group 14: Monomers and Polymerizable Esters

Acrylic acid (inhibited)
Acrylonitrile
Butadiene (inhibited)
Butyl acrylate (n, iso)
Ethyl acrylate (inhibited)
2-Ethylhexyl acrylate (inhibited)
Isodecyl acrylate (inhibited)*

FIG. 20−1 (Continued)

Isoprene (inhibited)
Methyl acrylate (inhibited)
Methyl methacrylate (inhibited)
σ−Propiolactone*
Styrene (inhibited)
Vinyl acetate (inhibited)
Vinyl chloride (inhibited)
Vinylidene chloride (inhibited)
Vinyl toluene

Group 15: Phenols

Carbolic oil
Creosote, coal tar*
Cresols
Nonylphenol
Phenol

Group 16: Alkylene Oxides

Ethylene oxide
Propylene oxide

Group 17: Cyanohydrins

Acetone cyanohydrin
Ethylene cyanohydrin

Group 18: Nitriles

Acetonitrile
Adiponitrile

Group 19: Ammonia

Ammonium hydroxide

Group 20: Halogens

Bromine
Chlorine

Group 21: Ethers

Diethyl ether (ethyl ether)
1,4-Dioxane
Isoprophyl ether*
Tetrahydrofuran

Group 22: Phosphorus, Elemental

Group 23: Sulfur, Molten

Group 24: Acid Anhydride

Acetic anhydride
Propionic anhydride

* Not presently included in CHRIS system.

TABLE 20–1 Chemical Incompatibilities

Chemical	Is Incompatible with
Acetic acid	Chromic acid, nitric acid, hydroxyl compounds, ethylene glycol, perchloric acid, peroxides, permanganates
Acetylene	Chlorine, bromine, copper, fluorine, silver, mercury
Acetone	Concentrated nitric and sulfuric acid mixtures
Alkali and alkaline earth metals (e.g., powdered aluminum or magnesium, calcium, lithium, sodium, potassium)	Water, carbon tetrachloride or other chloriniated hydrocarbons, carbon dioxide, halogens
Ammonia (anhydrous)	Mercury (e.g., in manometers), chlorine, calcium hypochlorite, iodine, bromine, hydrofluoric acid (anhydrous)
Ammonium nitrate	Acids, powdered metals, flammable liquids, chlorates, nitrates, sulfur, finely divided organic or combustile materials
Aniline	Nitric acid, hydrogen peroxide
Arsenical materials	Any reducing agent
Azides	Acids
Bromine	See Chlorine
Calcium oxide	Water
Carbon (activated)	Calcium hypochlorite, all oxidizing agents
Carbon tetrachloride	Sodium
Chlorates	Ammonium salts, acids, powdered metals, sulfur, finely divided organic or combustible materials
Chromic acid and chromium trioxide	Acetic acid, naphthalene, camphor, glycerol, alcohol, flammable liquids in general
Chlorine	Ammonia, acetylene, butadiene, butane, methane, propane (or other petroleum gases), hydrogen, sodium carbide, benzene, finely divided metals turpentine

TABLE 20-1 (Continued)

Chlorine dioxide	Ammonia, methane, phosphine, hydrogen sulfide
Copper	Acetylene, hydrogen peroxide
Cumene hydroperoxide	Acids (organic or inorganic)
Cyanides	Acids
Flammable liquids	Ammonium nitrate, chromatic acid, hydrogen peroxide, nitric acid, sodium peroxide, halogens
Fluorine	Everything
Hydrocarbons (e.g., butane, propane, benzene)	Fluorine, chlorine, bromine, chromic acid, sodium peroxide
Hydrocyanic acid	Nitric acid, alkali
Hydrofluoric acid (anhydrous)	Ammonia (aqueous or anhydrous)
Hydrogen peroxide	Copper, chromium, iron, most metals or their salts, alcohols, acetone, organic materials, aniline, nitromethane, combustible materials
Hydrogen sulfide	Fuming nitric acid, oxidizing gases
Hypochlorites	Acids, activated carbon
Iodine	Acetylene, ammonia (aqueous or anhydrous), hydrogen
Mercury	Acetylene, fulminic acid, ammonia
Nitrates	Sulfuric acid
Nitric acid (concentrated)	Acetic acid, aniline, chromic acid, hydrocyanic acid, hydrogen sulfide, flammable liquids, flammable gases, copper, brass, any heavy metals
Nitrates	Acids
Nitroparaffins	Inorganic bases, amines
Oxalic acid	Silver, mercury
Oxygen	Oils, grease, hydrogen, flammable liquids, solids, or gases
Perchloric acid	Acetic anhydride, bismuth and its alloys, alcohol, paper, wood, grease, oils
Peroxides, organic	Acids (organic or mineral), avoid friction, store cold
Phosphorus (white)	Air, oxygen, alkalis, reducing agents
Phosphorus pentoxide	Water

234

Potassium	Carbon tetrachloride, carbon dioxide, water
Potassium chlorate	Sulfuric and other acids
Potassium perchlorate (see also chlorates)	Sulfuric and other acids
Potassium permanganate	Glycerol, ethylene glycol, benzaldehyde, surfuric acid
Selenides	Reducing agents
Silver	Acetylene, oxalic acid, tartartic acid, ammonium compounds, fulmunic acid
Sodium	Carbon tetrachloride, carbon dioxide, water
Sodium nitrate	Ammonium nitrate and other ammonium salts
Sodium peroxide	Ethyl or methyl alcohol, glacial acetic acid, acetic anhydride, benzaldehyde, carbon disulfide, glycerin, ethylene glycol, ethyl acetate, methyl acetate, furfural
Sulfides	Acids
Sulfuric acid	Potassium chlorate, potassium perchlorate, potassium permanganate (similar compounds of light metals, such as sodium, lithium)
Tellurides	Reducing agents

SOURCE: *Prudent Practices for Handling Hazardous Chemicals in Laboratories*, National Research Council, Washington, D.C., 1981.

235

Reactivity Group No.	Reactivity Group Name	1	2	3	4	5	6	7	8	9	10	11	12	13	14	15
1	Acids, Mineral, Nonoxidizing	1														
2	Acids, Mineral, Oxidizing		2													
3	Acids, Organic	G H		3												
4	Alcohols and Glycols	H	H F	H P	4											
5	Aldehydes	H P	H P	H P		5										
6	Amides	H	H GT				6									
7	Amines, Aliphatic and Aromatic	H	H GT	H		H		7								
8	Azo Compounds, Diazo Compounds, and Hydrazines	H G	H GT	H G	H G	H			8							
9	Carbamates	H G	H GT						G H	9						
10	Caustics	H	H	H		H				H G	10					
11	Cyanides	GT GF	GT GF	GT GF					G			11				
12	Dithiocarbamates	H GF F	H GF F	H GF GT		GF GT		U	H G				12			
13	Esters	H	H F						H G		H			13		
14	Ethers	H	H F												14	
15	Fluorides, Inorganic	GT	GT	GT												15
16	Hydrocarbons, Aromatic		H F													
17	Halogenated Organics	H GT	H F GT					H GT	H G		H GF	H				
18	Isocyanates	H G	H F GT	H G	H P			H P	H G		H P G	H G	U			
19	Ketones	H	H F						H G		H	H				
20	Mercaptans and Other Organic Sulfides	GT GF	H F GT						H G							
21	Metals, Alkali and Alkaline Earth, Elemental	GF H F	GF H F	GF H F	GF H F	GF H F	GF H	GF H	GF H	GF H	GF H	GF H	GF GT H	GF H		
22	Metals, Other Elemental & Alloys as Powders, Vapors, or Sponges	GF H F	GF H F	GF							H F GT	U	GF H			
23	Metals, Other Elemental & Alloys as Sheets, Rods, Drops, Moldings, etc.	GF H F	GF H F								H F G					
24	Metals and Metal Compounds, Toxic	S	S	S			S	S			S					
25	Nitrides	GF H F	H F F	H GF	GF H F	GF H				U	H G	U	GF H	GF H	GF H	
26	Nitriles	H GT GF	H F GT	H								U				
27	Nitro Compounds, Organic		H F GT			H					H E					
28	Hydrocarbons, Aliphatic, Unsaturated	H	H F			H										
29	Hydrocarbons, Aliphatic, Saturated		H F													
30	Peroxides and Hydroperoxides, Organic	H G	H E		H F	H G		H GT	H F E	H F GT		H E GT	H F GT			
31	Phenols and Cresols	H	H F						H G							
32	Organophosphates, Phosphothioates, Phosphodithioates	H GT	H GT					U			H E					
33	Sulfides, Inorganic	GT GF	HF GT	GT		H			E	·						
34	Epoxides	H P	H P	H P	H P	U		H P	H P		H P	H P	U			
101	Combustible and Flammable Materials, Miscellaneous	H G	H F GT													
102	Explosives	H E	H E	H E				H E		H E				H E		
103	Polymerizable Compounds	P H	P H	P H				P H		P H	P H	U				
104	Oxidizing Agents, Strong	H GT		H GT	H F	H F	H F GT	H F GT	H E	H F GT	H E GT	H F GT	H F	H F		
105	Reducing Agents, Strong	H GF	H F GT	H GF	H GF F	GF H F	GF H	H GF	H G		H GT	H F				
106	Water and Mixtures Containing Water	H	H						G							
107	Water Reactive Substances	← EXTREMELY REACTIVE! →														
		1	2	3	4	5	6	7	8	9	10	11	12	13	14	15

FIG. 20–2: Hazardous waste compatibility chart

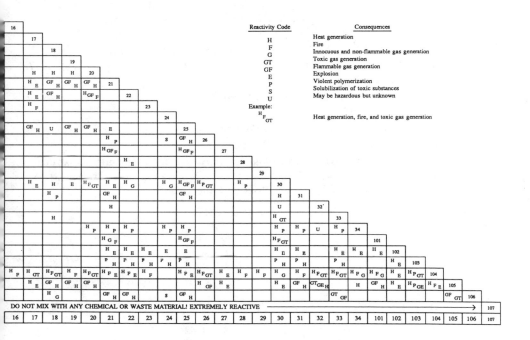

Source: H. K. Hatayama et al., EPA 600/2−80−076 "A Method for Determining the Compatibility of Hazardous Waste,"

IV CHEMICAL HYGIENE PLAN CONSIDERATIONS

The chemical hygiene plan should address chemical storage and handling requirements designed to avoid mixing incompatible chemicals. Incompatibility guides such as those included in this chapter should be incorporated into the hygiene plan. Periodic inspections to check for proper segregation of incompatibles should be performed.

RESOURCES

Hatayama, H. K., et al., "A Method for Determining the Compatibility of Hazardous Waste," EPA Document 600/2−80−076. Washington, DC: Government Printing Office, 1980.

Jackson, H. S., W. B. McCormack, G. S. Rondestvedt, K. C. Smeltz, and I. E. Viele "Safety in the Chemical Laboratory" *J. Chem. Educ.* 47:(3), A176 (March 1970).

Pepitone, D. A. (Ed.), *Safe Storage of Laboratory Chemicals*. New York: Wiley-Interscience, 1984.

Product Practices for Handling Hazardous Chemicals in Laboratories. Washington, DC: National Research Council, 1981.

U.S. Coast Guard, *CHRIS Hazardous Chemical Data*, Commandant Instruction M.16465.12A., 0−479−762:QL3, U.S. Department of Transportation. Washington, DC: Government Printing Office, 1985.

CHAPTER 21

BIOHAZARDS

I INTRODUCTION

Laboratory-associated infections, usually characterized by delayed onset, are less readily recognized than acute health effects associated with chemical exposure. Where laboratory animals, particularly rodents and primates, are used in research and testing laboratories, their presence may introduce a source of exposure to zoonotic diseases. Employees in in vitro testing laboratories may also be concerned with exposure to biohazards.

II ANIMAL DISEASES

Infection from animals can occur by numerous routes, including bites; scratches; shedding of infectious agents through feces, urine, or other excretions; coat shedding; and aerosols created by animal respiratory patterns. However, the most frequently encountered exposures involve skin accidentally punctured with infected needles, necropsy exposures from lacerations and splashes, and manual conjunctival exposures.

There are more than 150 recognized animal zoonoses. About 30 of them can be transmitted from laboratory animals to man. A study of workers at the National Animal Disease Center (NADC) reported 128 laboratory exposures to zoonotic organisms over 15 years. They resulted in 34 infections. The organisms the NADC study found to constitute the highest exposure risk were *Brucella* spp., *Mycobacteria* spp., and *Leptospira* spp. Rodents in particular may carry lymphocytic choriomeningitis (from a virus), leptospirosis or salmonellosis (from bacteria), *Escherichia coli* bacteria, and dermatosis (from fungus).

III CLASSIFICATION OF ETIOLOGIC AGENTS ON THE BASIS OF HAZARD

The scientific community has classified biological agents in their order of risk to humans. Each classification relates to a laboratory biosafety level, with safety equipment, procedures, and facility design combined to determine the minimum amount of protection required to preclude exposure. The following agent classification, as well as biosafety levels, is taken from the manual entitled *Biosafety in Microbiological and Biomedical Laboratories*, prepared by the Centers for Disease Control (CDC) in conjuntion with the National Institutes of Health (NIH) [1]:

Class 1. Agents of no or minimal hazard under ordinary conditions of handling.

Class 2. Agents of ordinary potential hazard. This class includes agents that may produce diseases of varying degrees of severity from accidental inoculation or injection or other means of cutaneous penetration but are contained by ordinary laboratory techniques.

Class 3. Agents involving special hazards, or agents derived from outside the United States, that require a federal permit for importation, unless they are specified for higher classification. This class includes pathogens that require special conditions for containment.

Class 4. Agents that require the most stringent conditions for their containment because they are extremely hazardous to laboratory personnel, or may cause serious epidemic diseases. This class includes Class 3 agents from outside the United States when they are employed in entomological experiments, or when other entomological experiments are conducted in the same laboratory area.

Class 5. Foreign animal pathogens.

Laboratories should consider these classifications and biosafety levels as recommendations. It is important to remember that all aspects of a biosafety level together work in reducing the exposure and escape of pathogenic microbes.

IV IMPLEMENTING BIOSAFETY PROGRAMS

A Control of Handling Procedures

Infection control plays a dual role in the laboratory. Techniques, procedural guidelines, engineering controls, and facility design are important in the protection of the worker as well as the experiment. Cross-contamination between animals caused by human error can be very costly and frustrating. The first task in controlling or ensuring against exposure to a biohazard is to recognize the hazard. The next task is to estimate how much concern exists for that hazard. The following subsections briefly explain the information and materials necessary to promote a good biohazard program.

1. Information Dissemination

Before a hazard can be controlled, or the necessary precautions taken in handling the agent, those at risk must know what pathogen they may possibly encounter. The international biohazard sign (Fig. 21−1) should be posted with care in any area where a known pathogen is in use. The principal investigator and/or the health and safety committee should review each program and identify all areas that should be posted. The sign should be removed after the experiment has been completed and the room has been decontaminated.

The laboratory should train its support personnel in all aspects of infection control. The training should emphasize the proper microbiological and animal-handling techniques necessary to control the transfer

FIG. 21−1:

of an infection. Good laboratory industrial hygiene practices are keys to protecting the worker and the experiment.

2. Facility Design

A poorly designed facility can contribute to the spread of microorganisms throughout a laboratory. Thus, the laboratory should review each project to assess the proper facility design necessary.

The following is a general summary of the protection recommended for each of the biosafety hazard levels. The actual criteria are contained in the CDC/NIH manual [1].

Biosafety Level 1   Basic (for agents of no known or minimal potential hazard; generally, Class 1 agents)

- Discretionary limited access to the laboratory when experiments are in progress.
- Surfaces are designed to be easily cleaned. Bench tops are impervious to water and resistant to acid, alkali, organic solvents, and moderate heat.
- Work surfaces are decontaminated daily and after spills.
- Contaminated liquid or solid wastes are decontaminated before disposal.
- Mechanical pipettes are used.
- No eating, drinking, or smoking is permitted.
- Hands must be washed after handling viable materials and animals and before leaving the laboratory.
- Procedures are performed in a manner most likely not to produce aerosols.
- Laboratory coats, gowns, or uniforms are worn.
- Insect control and rodent control programs are in effect.

Biosafety Level 2   Basic (for work involving agents of moderate potential hazard; generally, Class 2 agents)

Same guidelines as Biosafety Level 1, but also:

- Access to the laboratory is limited when work is being conducted.
- Personnel are trained in handling pathogenic agents and are directed by compenent scientists.
- Biological safety cabinets or other physical containment equipment is used when aerosols are generated.

- A universal hazard warning sign is placed on the door of the laboratory, identifying the infectious agent, giving precautions that are to be taken before entry, and listing the name and telephone number of the person(s) to be contacted in the event of an emergency.
- Gloves are worn to avoid skin contamination with infectious materials.
- Needle-locking syringes or disposable syringe—needle units are used. Needles are not broken, and needles and syringes are autoclaved before disposal, in an impervious container.
- Spills and exposures are reported. Medical treatment is provided when necessary, and written records are maintained.
- A biosafety manual is prepared.

Biosafety Level 3   Containment (work with indigenous or exotic agents that may cause a serious or potentially lethal disease as a result of exposure from inhalation; generally, Class 3 agents)
Same guidelines as Biosafety Level 2, but also:

- Laboratory doors are closed when experiments are in progress.
- Decontamination occurs at another site, away from the laboratory.
- Access is highly restrictive, including a double set of doors separating the laboratory from other areas.
- Each laboratory is equipped with a sink.
- The universal biohazard symbol is posted on laboratory doors.
- Infectious work is done in a biological safety cabinet.
- Disposable plastic-backed matting is used on nonperforated work surfaces within biological safety cabinets.
- Respirators or surgical masks are worn in infected animal rooms.
- Vaccum lines are adapted with high efficiency particulate air (HEPA) filters and liquid disinfectant traps.
- Exhaust from biological safety cabinets is HEPA-filtered and discharged directly to the outside.
- Ducted exhaust air ventilation systems are used. Exhaust airflow is not recirculated.
- Baseline serum samples are collected for all at-risk personnel.

Biosafety Level 4   Maximum containment (for work that involves dangerous and exotic agents that present a risk of life-threatening disease; generally, Class 4 agents)
Same guidelines as Biosafety Level 3, but also:

- Work is conducted in a Class III biological safety cabinet or in a Class I or II unit with one-piece, positive-pressure suits.
- Access is under strict control.
- Showering is required on entry and exit.
- Transfer of materials requires nonbreakable secondary containment.
- Delivery and shipment of supplies and materials is effected through a double-door autoclave, fumigation chamber, or ventilated airlock.
- Facility design is such that interior walls, floors, and ceiling form a tight seal, and bench tops have a minimal number of seams.
- Exhaust and effluents are decontaminated before being released from the facility.

Additional general safety practices for handling agents include:

- Never put a pipette to the mouth.
- Do not eat, drink, smoke, or chew gum in the laboratory.
- Label all rooms or equipment with the universal biohazard symbol when biohazards are used. Be sure to remove all labels upon decontamination.
- Use needle-locking (i.e., Luer-lok) syringes or disposable syringe–needle units.
- Place used syringes and needles in an impervious container. Do *not* clip or shear needles after use.
- Equip centrifuges with trunnion cups with germicidal solution to prevent the spread of contamination from broken vials.
- Wash hands thoroughly after completing experiments and before leaving the laboratory.
- Place contaminated pipettes in disinfectant solution after use and then autoclave them.
- Use only cotton-plugged pipettes.
- Place wastes that have been contaminated by an infectious agent in a sealed, labeled plastic bag, to be autoclaved later at the appropriate temperature and pressure.
- Monitor autoclaves for effectiveness with biological indicators.
- Work with infectious material in a biological safety cabinet that will adequately contain the hazard.
- Wear appropriate personal protective equipment for the biohazard (i.e., gloves, coat, suit, mask, and/or respirator).

In addition to the major design features of a laboratory, the following features will enhance the control of microorganisms:

- Room pressure differentials — any laboratory that holds an infectious agent or an animal experimentally inoculated with an infectious agent, or an animal room, should be under negative pressure with respect to the outer areas.
- Air coming into or out of a biohazard area should be made to pass through a HEPA filter.
- A biohazard laboratory is a restricted-access area.
- Animal rooms or biohazard areas should be located away from the stream of general traffic movement.
- Where possible, ultraviolet lights should be used to control the spread of microorganisms as an overnight decontamination process.

3.   Human Fluid Samples

Laboratories that use cultures and human samples as part of their experimental protocol must recognize the potential hazards associated with such work. Although some agents present a more definite hazard (i.e., blood samples and lymphoid cell lines), the actual hazards are not clearly recognized. Protection from unknown or nondefined hazards should be provided for. The American Industrial Hygiene Association (AIHA) has described the containment requirements for some cell/tissue cultures as follows:

Minimal containment

- primary peripheral lymphocytes without passage, and
- primary explants of febroblasts from benign tissues, up to and including first passage.

Low containment

All other cell types except those containing

- viral agents of Class 3 or 4, and
- agents described in moderate containment.

Moderate containment

- cells containing Herpes, Adenovirus-Simian Virus-10 hybrid viruses, human hepatitis-associated virus;
- unknown cultures derived from malignant primate tissues; and
- moderate or high risk National Cancer Institute (NCI) oncogenic viruses.

High containment

- cells containing specific viral agents of CDC Class 4 or oncogenic viruses considered by NCI as moderate risk.

The Occupational Safety and Health Administration (OSHA) and CDC have started to develop regulations that govern the manipulation of cells, tissue culture, or sample containing Hepatitis B or human immuno-deficiency virus (HIV) associated with AIDS viruses. OSHA has issued an Advanced Notice of Proposed Rulemaking (ANPR) for handling of specimens contaminated with Hepatitis B virus (HBV), HIV, and cyto-megalovirus (CMV) with respect to their potential threat to pregnant women. The ANPR stresses prevention of transmission of these viruses. Questions for public comment include methods of controlling exposures (engineering controls), personal protective equipment, administrative programs, management of injuries (e.g., needlesticks, cuts, slashes), medical surveillance, training and education, and the necessity of a generic standard to cover similar blood-borne diseases. Laboratories should follow these public hearings and any resultant regulations or guidelines that emerge.

The CDC has issued [2] specifically directed at health care workers (i.e., persons having contact with patients[1] *blood* or other body fluids) who can be in contact with the HIV virus. The recommendations provide control practices for handling HIV in invasive procedures, dentistry, autopsies, dialysis, and laboratories. The CDC recommendation for laboratories are as follows:

1. All specimens of blood and body fluids should be put in well-constructed containers having secure lids, to prevent leaking during transport. Care should be taken when collecting each specimen to avoid contaminating the outside of the container and of the laboratory form accompanying the specimen.

2. All persons processing blood and body fluid specimen (e.g., removing tops from vacuum tubes) should wear gloves. Masks and protective eyewear should be worn if mucous membrane contact with blood or body fluids is anticipated. Gloves should be changed and hands washed after completion of specimen processing.

3. For routine procedures, such as histologic and pathologic studies or microbiologic culturing, a biological safety cabinet is (Class I or II) should be used whenever procedures are conducted that have a high potential for generating droplets. (e.g., blending, sonicating, vigorous mixing).

4. Mechanical pipetting devices should be used for manipulating all liquids in the laboratory. Mouth pipetting must not be done.

5. Use of needles and syringes should be limited to situations in which there is no alternative, and the recommendations for preventing injuries with needles outlined above should be followed.

6. Laboratory work surfaces should be decontaminated with an appropriate chemical germicide after a spill of blood or other body fluids and when work activities are completed.

7. Contaminated materials used in laboratory tests should be decontaminated before reprocessing or be placed in bags and disposed of in accordance with institutional policies for disposal of infective waste.

8. Scientific equipment that has been contaminated with blood or other body fluids should be decontaminated and cleaned before being repaired in the laboratory or transported to the manufacturer.

9. All persons should wash their hands after completing laboratory activities and should remove protective clothing before leaving the laboratory.

Implementation of universal blood and body fluid precautions for *all* patients eliminates the need for warning labels on specimens, since blood and other body fluids from all patients should be considered infective.

Another critical section of the recommendations provides guidance on sterilization and disinfection. The document states that chemical germicides registered with the U.S. Environmental Protection Agency (EPA) as sterilants are effective in inactivating HIV depending on the contact time and concentration. A solution of sodium hypochlorite (bleach) prepared fresh daily in concentrations of 1:10–1:100 will also disinfect surfaces. If laboratories are using human blood or body fluids, these CDC recommendations must be incorporated into the standard operating procedures.

Recombinant DNA products may also present a biohazard risk to those who use them. Most DNA-associated experimentation also presents chemical and radioactive concerns. This type of work should be treated with cautious measures that protect against all hazards. Work should be conducted in a Biosafety Level 3 area.

4. Engineering Controls

Laboratories use biological safety cabinets (BSCs) to confine an agent, as well as to provide protection to the individual using that agent. Chapter 12 presents a complete discussion of biological safety cabinets. See Biosafety Levels above and Chapters 8 and 9 for guidance on general ventilation systems.

5. *Wastes*

Biological wastes, like chemical and radioactive wastes, require an effective waste management program and a review of the specific regulations pertaining to the individual laboratory. Such a program must address predisposal decontamination of the biological agent. Chapter 24 details the various decontamination procedures that may be included in a biological waste management program.

V CHEMICAL HYGIENE PLAN CONSIDERATIONS

The chemical hygiene plan should address biohazards if such are present in the laboratory. The plan should indicate that the recommendations for labeling, handling, and so on, found in *Biosafety in Microbiological and Biomedical Laboratories* [1] are to be followed as well as the following specific requirements:

1. Class 3 and 4 infectious agents are to be weighed, handled, and administered in facilities with primary and secondary containment including glove boxes.

2. When handling infectious agents, the following precautions are to be followed:

- Syringes shall be needle-locking (e.g., Luer-Lok) or disposable syringe−needle units. After use, intact syringe and needle shall be placed in a puncture-resistant container and decontaminated. Needles shall *not* be clipped or sheared after use.
- Centrifuges shall be equipped with trunnion cups with germicidal solution to prevent the spread of contamination from broken vials.
- Contaminated pipettes shall be placed in a disinfectant solution and autoclaved after use. Pipettes used outside of glove boxes shall be plugged with cotton.
- When removing a syringe and needle from a rubber-stoppered bottle containing infectious agents, an alcohol pledget around the stopper and needle shall be used.

3. All solid and liquid wastes contaminated by an infectious agent are to be placed in a sealed, labeled plastic bag and autoclaved immediately.

4. Autoclaves are to be posted with the maximum acceptable pressure and date of certification. The effectiveness of the autoclave is to be monitored periodically with biological indicators.

The following general requirements relating to biohazard control as well as exposure to chemicals should be observed:

- No eating, smoking, drinking, or applying cosmetics is to be done in the laboratory.
- Laboratory clothing is restricted to the areas to which the garments are assigned. Laboratory coats are not allowed in the eating areas.
- Hands are to be washed thoroughly before leaving the laboratory.
- No mouth pipetting.

REFERENCES

1. Centers for Disease Control−National Institutes of Health, *Biosafety in Microbiological and Biomedical Laboratories*, 1st ed. Washington, DC: Department of Health and Human Services, 1984.
2. Centers for Disease Control, "Recommendations for Precaution of HIV Transmission in Health-Care Settings," *Morb. Mortal. Wkly. Rep.*, 36:25 (Aug. 21, 1987).

RESOURCES

AIHA Biohazards Committee, *Biohazards Reference Manual*. Akron, OH: American Industrial Hygiene Association, 1985.

Collins, C. H., *Laboratory-Acquired Infections*. Boston: Butterworths, 1983.

Fuscaldo, A. A., B. J. Erlich, and B. Hindman (Eds.), *Laboratory Safety — Theory and Practice*. New York: Academic Press, 1980.

Miller, C. D., J. R. Songer, and J. F. Sullivan, "A Twenty-Five Year Review of Laboratory-Acquired Human Infections at the National Animal Disease Center", *Am. Ind. Hygiene Associ. J.* 48:3, 271−275 (1987).

National Institutes of Health, *Biohazards Safety Guide*. Washington, DC: Department of Health, Education and Welfare, 1974.

CHAPTER 22

RADIATION

I INTRODUCTION

The ionizing radiation that is emitted by unstable elements can be a very useful tool in the laboratory. In some laboratories, radiation is commonly used as a tracer for chemicals in biological or chemical research. In this procedure, the laboratory technician substitutes a radioactive atom for its stable counterpart in a chemical compound. This permits the tracing of quantities of the chemical to their final destination. For example, to determine the location and quantity of glucose in a biological system, the technician can use a compound labeled with the isotope carbon-14 (^{14}C). The "tagged" glucose will mix completely with the native glucose pools, since the two are indistinguishable. With this procedure, the technician can easily determine the quantity and location of the glucose by performing a radioactive assay of various samples. For the proper use of tracers, one must know the physical and chemical nature of the compound used.

The applications of laser, microwave, and ultrasound technologies are necessary in many laboratories, and when using radioactive materials, one must take proper precautions to prevent the contamination of equipment, working area, and, most important of all, people. Poor laboratory safety skills can result in severe problems, as well as repercussions for the institution or facility in the form of fines, lawsuits, or even loss of licenses. This chapter presents means of developing a safe working environment by discussing laboratory safety precautions and what a laboratory needs to implement a radiation safety protection program. A program for handling nonionizing radiation is offered, as well.

II IMPLEMENTATION OF A RADIATION SAFETY PROGRAM

Before a facility can begin receiving radiochemicals, its radiation safety officer (RSO) should be notified. The RSO needs to know:

- from whom the radiochemicals are to be obtained,
- where they are to be used,
- which ones are to be used, and
- how much will be needed.

This information gives the RSO the basis for determining which federal regulations apply to the situation and whether the licenses presently held will allow the operation.

The federal government controls any laboratory involvement with radioactive materials and elements. In fact, the executive departments and agencies of the federal government have established general and permanent rules for the handling and control of radioactive materials. These have been published in the Code of Federal Regulations (CFR) and the Federal Register (FR). Title 10 of the CFR, which the Nuclear Regulatory Commission (NRC) oversees, contains all necessary information on the use of radiation. The NRC attempts to maintain tight and restrictive control on the holders and users of radioactive material. The NRC performs periodic inspections to ensure compliance with the regulations by any and all licensed holders and users of radioactive material.

Any individual who becomes involved with radioactive materials should also become familiar with Title 10, especially Part 20 (i.e., 10 CFR 20). Each laboratory that uses radioactive materials should post 10 CFR 20, in a conspicuous area. The RSO will supply any other regulations that should be posted in the laboratory as well.

It is the policy of most laboratories to urge their employees not to expose themselves to any more radiation than is necessary. This is known as the ALARA concept — as low as reasonably achievable. To check the radiation level to which a worker may have been exposed, several monitoring methods are available, as listed below. The RSO should advise on the efficacy of each method.

- Designation of a control area where the workers can dress in the appropriate manner and remove protective clothing upon completion of their work.

- Use of a dosimeter to check the radiation exposure of each worker. Film badges or a pocket ionization chambers are common forms of dosimeter. These devices allow the RSO to quantitatively determine the amount of radiation to which the worker has been exposed.
- A wipe test program to check for area contamination. In this method, the RSO or a deputy takes periodic wipes of the laboratory and counter area using special paper or any 2-in. filter paper. The same person then checks the wipes for contamination with a radiation detector.

The radiation safety officer can help a laboratory to set up an effective radiation protection program. Such a program should consist of all the safety procedures discussed previously, as well as the following:

1. administration of a personnel monitoring service with records that are kept up-to-date;
2. performance of routine radiation contamination surveys in the laboratory and counter areas, with complete records to be analyzed for trends;
3. frequent maintenance and calibration of instrumentation, basis with results recorded;
4. recording of all radioactive materials and the date received;
5. provision of protective clothing labels, absorbent paper, and other equipment needed to maintain a healthful working atmosphere, as described later; and
6. control of radioactive waste, with efforts made to minimize the amount of waste produced without compromising safety.

The disposal of radioactive waste is very costly. Therefore, a laboratory should keep accurate records of the amount of waste generated.

To ensure an atmosphere that is conducive to the safe handling and use of radiochemicals, a laboratory should establish certain regulations, including the following:

1. Do not eat, drink, smoke, or chew gum, or use cosmetics in the laboratory.
2. Do not perform pipetting or any similar operation by mouth suction.
3. Before leaving the working area, wash hands and check for contamination with the gloves on, then off.

4. Perform routine surveys with an appropriate instrument, when personal contamination is suspected.
5. Perform all work with rubber gloves.
6. Wear some form of personal monitoring device (e.g., a dosimeter).
7. Pour liquids contaminated by radiation into containers and mark them with a label that indicates the type and amount of radioactive substance.
8. Dispose solid wastes in a "contaminated" waste can.
9. In the event of spillage:
   (a) blot the liquid with tissue or absorbent paper, and use rubber gloves or gloves recommended for handling a particular chemical;
   (b) place all disposable materials contaminated by the spill and those used in the cleaning process in a trash can marked "contaminated"; and
   (c) clearly mark the area with chalk or tape borders and designate the type of spill.
10. Do not wash any equipment in a public water sewage system unless previously checked for contamination. Provide a sink for contaminants.
11. Keep all radioactive materials within the radiochemical laboratory in designated areas.
12. Report all wounds, spills, or emergencies to the RSO immediately.
13. Complete all records of activities before leaving the laboratory.

The radiochemical laboratory consists of two specific areas: (1) the work area or laboratory proper, and (2) the room or area for performing the quantitative measurement of radioactivity, commonly referred to as the counter room. It is especially important that the laboratory staff keep the counter room free of contamination, to provide a low background of radioactivity and to maintain healthy working conditions.

The laboratory staff must perform all preparative work for radiochemical experiments in the work area — never in the counter room — with all work restricted to individual working areas. A metal tray lined with an absorbent paper might be designated as an "area." On such a tray, the paper has two layers: the upper is highly absorbent, and the lower is a plastic backing to prevent liquid from soaking through to the tray or table surface. The technician should try to perform all work within the confines of the tray. As an added precaution, the table top should be covered with an absorbent paper. Then, should a spill occur, the contaminated paper can be removed and replaced easily.

The laboratory staff should place equipment that becomes contaminated through normal use in a designated area, or container, with tissue or absorbent paper. They should dispose of solid wastes and pour liquid wastes into containers marked "radioactive," with the type and amount of radioactive chemical noted.

The laboratory staff should always wear rubber gloves, or gloves recommended for use in handling a particular chemical when working with radiochemicals. This includes handling any contaminants, stock solution containers, and waste containers, and the disposal of absorbent paper or tissues.

At the conclusion of their work, the staff should wash their hands with their gloves still on, making sure not to contaminate the faucet appliances. Also, they should check their gloves for contamination while still on; if free of contamination, they may take them off, and then wash their hands. If their gloves are still contaminated, they should wash them again, check for contamination, and then remove them, turning them inside out to prevent the ungloved hand from touching the outside contaminated surface. Finally, they should dispose of the gloves in the waste can marked "radioactive."

One should not wear gloves in the counter area or room, for it is imperative that the area be kept free from radiation contamination at any level. One should also wear a disposable laboratory coat to protect street clothing, with the pockets taped shut to prevent the unnoticed collection of radioactive materials.

Special precautions in handling radioactive materials may be necessary, depending upon the radioisotope used. For example, phosphorus-32 (^{32}P) emits a highly energetic beta particle and, therefore, requires special handling, shielding, and dosimetry. Laboratory workers should handle only that amount of ^{32}P necessary to perform the experiment and should work from behind a polycarbonate or plastic shield. Doses to the fingers are of concern when handling ^{32}P, so a worker should wear a finger badge. The proper shielding for storage of ^{32}P solutions consists of a polycarbonate cylinder inside a lead pig. Before working with isotopes other than ^{14}C and tritium (^{3}H), one should consider all information on special precautions. Table 22−1 presents specific information on some of the more common laboratory isotopes. Whenever handling isotopes, one should review protective measures for the specific radionuclide being used.

At the conclusion of the laboratory experiment or working period, employees should survey their hands, feet, and clothing, taking to use the correct type of detector for the radiation being emitted. Finally, the work area should be surveyed to identify spills that were not observed during the work period.

TABLE 22–1 Specific Information on Some Common Laboratory Isotopes

Property	Americium-241	Tritium	Iodine-125	Carbon-14	Chromium-51	Sulfur-35	Phosphorus-32
Emission	Alpha	Low energy beta	Low energy beta, Low energy gamma	Low energy beta	Low energy beta	Low energy beta	High energy beta
Energy (MeV)	5.48	0.018	0.030, 0.035	0.159	0.315, 0.320	0.167	1.7
Half-life	458 years	12.3 years	60 days	5730 years	27.7 days	87 days	14.3 days
Critical organ	None	Whole body	Thyroid	Fat	Whole body	Testis	Bone
Shielding	Skin, paper, lead, 0.002 cm aluminum	None needed	Lucite and lead 1/16[a]	1–5 mm lucite	Lucite and lead	Lucite and lead	1/4–1/2 in. lucite
Half-value layer[a] (HLV) Thickness lead	—	—	0.003 mm	—	0.2 cm	—	—
Protective equipment	None	Safety glasses Gloves Laboratory coat	Safety glasses Gloves Laboratory coat	Safety glasses Gloves Laboratory coat	Safety glasses Gloves Laboratory coat	Safety glasses Gloves Laboratory coat	Safety goggles Gloves Laboratory coat
Monitoring Whole body	None	None	X	None	X	None	X
Extremity	None	None		None	X	None	X
Bioassays	Urinalysis	Urinalysis	Urinalysis	Urinalysis	Urinalysis	Urinalysis	Urinalysis
Survey techniques	Geiger counter wipe test	Wipe tests	Thyroid scan Geiger counter, gamma counter (NaI "eye" crystal), wipe test	Geiger counter, wipe test	Geiger counter, gamma counter, wipe test	Geiger counter, gamma counter wipe test	Geiger counter gamma counter, wipe test
Labeling concentration	0.1 μCi	1 mCi	1 μCi	100 μCi	1 mCi	100 μCi	10 μCi
Posting concentration	0.1 μCi	10 mCi	10 μCi	1 mCi	10 mCi	1 mCi	100 μCi
Package leaktest	—	10 mCi	10 mCi	10 mCi	0.1 μCi	0.1 μCi	0.1 μCi

[a] Amount of lead that will diminish radiation by 50%

III EQUIPMENT

The type of equipment needed to start up a radiochemical laboratory depends on the radiochemicals that are to be used. Laundry services for contaminated clothing should also be made available. The radiation safety officer will advise on what equipment is needed and possible suppliers. (The RSO should supply the badges.) The following list indicates the types of equipment needed in a radiochemical laboratory:

beta/gamma survey meter;

^{14}C contamination monitor;

tritium surface contamination monitor;

direct-reading dosimeter;

dosimeter charger;

polyethylene bags;

radiation warning materials: warning signs, labels, tags, and "radioactive material" tape;

Rad-Con® decontaminant spray or foam for skin;

Isoclean® for equipment;

Decon® swipes;

absorbent paper;

protective trays;

rubber gloves;

disposable laboratory coats;

contaminated waste canisters; and

Kim-Wipe® tissues.

The next section presents a partial listing of suppliers of this equipment.

IV SUPPLIERS OF ADDITIONAL ITEMS

The following manufacturers, analytical laboratories, and equipment suppliers can identify the products and services necessary for implementing a radiation protection program.

Personnel monitoring devices

• Harshaw, Solon, OH

- Xetex Inc., Mountain View, CA
- Victoreen, Inc., Cleveland, OH

Portable radiation detectors

- Ludlum Measurements, Inc., Sweetwater, TX
- Eberline, Albuquerque, NM
- Xetex Inc., Mountain View, CA
- Victoreen, Inc. Cleveland, OH

Radiation detection instrumentation

- Nuclear Data Inc, Schaumburg, IL
- Harshaw, Solon, OH
- Xetex Inc. Mountain View, CA
- Tennelec, Oak Ridge, TN
- Canberra, Meriden, CT
- EG&G Ortec, Oak Ridge, TN
- Eberline, Albuquerque, NM

V NONIONIZING RADIATION

Nonionizing radiation does not possess sufficient energy to displace electrons that are bound to atoms. However, nonionizing radiation can damage atoms. In a laboratory environment, the most common forms of nonionizing radiation encountered are light, ultraviolet, lasers, and microwaves, all part of the electromagnetic spectrum. Another form of nonionizing radiation is ultrasound, which is not part of the electromagnetic spectrum.

This section presents the general principles that must be made a part of any safety program that controls the use of nonionizing radiation.

A Lasers

1 Recognition

The acronym LASER stands for "light amplification by the stimulated emission of radiation" Lasers are devices that produce light at very specific frequencies of the electromagnetic spectrum. The frequency of a laser depends on the type of material that is stimulated. The properties of lasers are similar to those of the other members of the electromagnetic spectrum. However, lasers can achieve great power densities which, along with operating at a single wavelength, has made them indispensable in today's marketplace.

Examples of types and uses of lasers in the medical field are:

- argon lasers, which operate in the 458- to 515-nm wavelength range and are used for photocoagulation;
- carbon dioxide lasers, which operates at a wavelength of 10.6 µm and are used in various surgical techniques;
- neodymium glass lasers made of yttrium–aluminum–gournet (YAG), which operate at a wavelength of 106 µm and are used in various surgical procedures; and
- ruby lasers, which operate at a wavelength of 694.3 nm, and are also used for photocoagulation.

The U.S. Food and Drug Administration (FDA) has stipulated that all manufacturers of lasers must meet the agency's performance standards. The standards created by the FDA divide laser products into four separate classes, based on the biological effect produced by the laser and the intensity of the radiation in the laser beam.

- Class I lasers produce radiation that causes no biological damage. The continuous output of Class I lasers is not more than 0.39 µw.
- Class II lasers produce radiation that can cause eye damage if exposures are direct and prolonged. The continuous output of a Class II laser is not more than 1 mW. These lasers operate in:

  (a) visible (400–700 nm) and continuous wave (CW) bands, and can emit a power exceeding P_{exempt} for the classification duration (0.4 mW for $T_{max} > 10^4$ s), but not exceeding 1 mW; and

  (b) visible (400–700 nm), repetitively pulsed, or scanning modes, which may be evaluated by specifying P_{exempt} at a point 10 cm from the exit port of the laser system. Such laser devices can emit a power that exceeds the appropriate P_{exempt} for the classification duration, but not exceeding P_{exempt} for a 0.25-second exposure.

- Class III lasers emit radiation that is powerful enough to damage skin tissue from direct or indirect exposures off of shiny surfaces for a short duration. The continuous output of Class III lasers is not more than 500 mW. These medium power laser devices produce radiation in:

   (a) the infrared (1.4 μm−1 mm) and the ultraviolet (200−400 nm) bands and can emit power in excess of P_{exempt} for the classification duration, but they cannot emit:

     (i) an average radiant power in excess of 0.5 W, for T_{max} greater than 0.25 second or

     (ii) a radiant exposure of 10 J cm^{-2} within an exposure duration of 0.25 second or less;

   (b) visible (400−700 nm) band, CW, or repetitive pulses, modes that produce a radiant power in excess of P_{exempt} for a 0.25-second exposure (1 mW for a CW laser), but they cannot emit an average radiant power of 0.5 W for T_{max} greater than 0.25 second;

   (c) visible and near-infrared (400−1400 nm) bands, and emit a radiant energy in excess of Q_{exempt}, but they cannot emit a radiant exposure that exceeds either 10 J cm^{-2} or that required to produce a hazardous diffuse reflection; and

   (d) near-infrared (700−1400 nm) CW band, CW, or they and repetitively pulsed modes emit tiny amounts of power in excess of P_{exempt} for the classification duration, but they cannot emit an average power of 0.5 W or greater for periods in excess of 0.25 second.

- Class IV lasers emit extremely powerful radiation, which can cause damage to tissue when exposures are short and the beam is direct, reflected, or diffused. The continuous output of a Class IV laser is more than 500 mW. These high power laser devices operate in:

   (a) ultraviolet (200−400 nm) and infrared (1.4 54m−1 mm) bands, and emit an average power in excess of 0.5 W for periods greater than 0.25 second or a radiant exposure of 10 J cm^{-2} within an exposure duration of 0.25 second or less; and

   (b) visible (400−700 nm) and near-infrared (700−1400 nm) bands, and emit an average power of 0.5 W or greater, for periods greater than 0.25 second, or a radiant exposure in excess of either 10 J cm^{-2}, or that required to produce a hazardous diffuse reflection.

    The only Class IV lasers that may be used in medicine are completely enclosed, so that accidental exposure to personnel cannot occur. Therefore, safety policies and procedures are not applicable.

    A facility should list sources of nonionizing radiation in its hazardous materials checklist which is a part of the procedure for recognizing hazardous materials (see Chapter 1). Before a laser is purchased, the facility should notify its health and safety officer (HSO), to permit the HSO to determine whether its use by laboratory personnel would be safe. Figure 22–1 is an example of a pre-purchasing form a user of a laser, or similar device, must forward to the HSO.

2 Evaluation

Typically, lasers require very little monitoring. The manufacturer designates both the power level and the wavelength. With this information, the facility can classify the laser into one of the four groups (I–IV) discussed above. Then, the facility can apply the controls based on the designated classification. The American Conference of Governmental Industrial Hygienists (ACGIH) lists the threshold limit values (TLVs) for lasers based on their exposure duration, radiation exposure, irradiance, and wavelength. The HSO can use the laser survey form shown in Figure 22–2 to help gather this information and apply the proper controls.

3 Control

Most control measures depend upon a laser's classification (I–V). In general, a Class I exempt laser device in considered to be incapable of producing damaging radiation levels. It is, therefore, exempt from any control measures or other forms of surveillance.

A Class II, lower power laser device may be viewed directly. However, it must bear a cautionary label warning against continuous intrabeam viewing.

A Class III, medium power laser device requires control measures that will preclude viewing of the beam directly.

A Class IV, high power laser device requires control measures that will preclude exposure of eyes as well as skin to the direct and diffusely reflected beam.

Class V lasers are either Class II, Class III, or Class IV lasers that have been contained in a protective housing and are operated so that they are incapable of emitting hazardous radiation from the enclosure. A facility must both install and maintain a stringent control system before any laser system can qualify for this level of classification.

1. *Class IV: Specific Precautions for High Power Laser Installation*

Pulsed Class IV visible and near-infrared lasers are hazardous to the eye from direct beam viewing and from specular and diffuse reflections

FIGURE 22−1: Sample facilities preparedness and procedures statement form.

Instrument name: _____

Manufacturer: _____

Laser Class   I   II   III   IV   V   Laser type: _____

Yes   No   Is instrument designed to be portable?

Manufacturer's suggestions for facility safeguards:	Additional facility safeguards deemed necessary:

Manufacturer's suggestions for safety procedures & PPE:	Additional safety procedures and Personal Protective Equipment (PPE) deemed necessary:
Protective eyewear	Training session for all employees Occupational vision program

The safety needs to protect employees from the potential hazards involved when using lasers have been evaluated, all parties agree on control measures indicated, training is complete, and work can proceed.

M.D. Responsible Physician	Safety Officer	Industrial Hygienist

of the laser beam. They are also generally hazardous to the skin. Safety precautions associated with high power lasers generally consist of using:

- door interlocks to prevent exposure to unauthorized or transient personnel entering the room,
- baffles to terminate the primary and secondary beams, and
- safety eyewear by personnel within the interlocked facility.

FIGURE 22–2

LASER SURVEY

| BUILDING NUMBER | ROOM NUMBER ORGANIZATION | PHONE |

OPERATORS

| NAME | SS | DATE OF PHYSICAL |

| HAZARD CONTROL EVALUATION | YES | NO |

1. Are laser warning signs displayed?

2. Is area secured or have limited access?

3. Is beam termination adequate?

4. Are laser safety glasses available?

5. Are laser safety glasses identified?

6. Is an SOP available?

7. Are personnel aware of the laser health hazards?

8. Is viewing of the beam with optical instruments performed?

9. Are precautions for toxic gases, fumes, or projectiles adequate?

10. Others?

11. Training

12.

13.

14.

FIGURE 22-2   (Continued)

EQUIPMENT IDENTIFICATION

Laser type:	Wavelength:	
Manufacturer:	Model no:	Serial no:
Beam diameter:	Beam divergence:	
Output power/energy:	Pulse duration:	Pulse repetition:

Safe eye exposure distance (SEED):

How is laser employed (alignment, scanning, airborne)?

PROTECTIVE EYEWEAR IDENTIFICATION

Protective glasses required:	OD of	@		
On hand:	OD of	@	Mfr/No	*VLT
Other suitable:	OD of	@	Mfr/No	*VLT
	OD of	@	Mfr/No	*VLT

* Visible light transmission

REMARKS

SURVEYED BY: DATE:

(a) A facility must deny access to unauthorized or transient personnel while the laser is capable of operating through the use of safety interlocks, or similar devices, at the entrance of the laser facility.

(b) Manufacturers should design laser electronic firing systems for pulsed lasers to preclude the accidental pulsing of a stored charge. For this purpose, the firing circuit design should incorporate a "fail-safe" system.

(c) The safety procedure for pulsed lasers should include use of an

alarm system, including a muted sound and/or flashing lights (visible through protective eyewear), and a countdown routine. This procedure should be initiated once the laser begins to charge.
(d) Good room illumination is important in areas where laser eye protection is required. Light colored, diffuse surfaces in the room help achieve this condition.
(e) Class IV pulsed ultraviolet, infrared, and all CW lasers are a potential fire and skin hazard.

2. *Specific Precautions Applicable to Class III Medium Power, CW, or Pulsed Laser Systems*

These lasers are potentially hazardous if the direct beam is viewed by the unprotected eye (intrabeam viewing). Care is required to prevent direct beam viewing and to control specular reflections.

(a) General safety precautions should be observed (e.g., not allowing the laser beam to be aimed at specular surfaces).
(b) The laser should be operated only in a well-controlled area.
(c) The laser beam should be terminated, where feasible, at the end of its useful beam path by interposing a material that is diffuse and of such color, or reflectivity, as to make positioning possible by minimizing the reflection.
(d) Specularly reflective material should be eliminated from the beam area, and good housekeeping should be maintained.
(e) Eye protection is required if direct beam (or specularly reflected beam) intraviewing is conceivable.

3. *Specific Precautions Applicable to Class II Low Power Visible Lasers*

Precautions are required only to prevent continuous staring into the direct beam; momentary (quarter-second) exposure, as would occur in an unintentional viewing situation, is not considered to be hazardous.

(a) The laser beam should not be purposefully directed toward the eye of any person for exposure durations that would be hazardous.
(b) A warning label reading: "CAUTION. DO NOT STARE INTO LASER BEAM" should be placed in a conspicuous location on the exterior housing of the laser.
(c) Scanning lasers, which must be scanning to meet the requirements of a Class II laser, should be designed to prevent laser emission when the scanning ceases.

Placarding of potentially hazardous areas should be accomplished in accordance with local standard operating procedures. Examples of signs are shown in Figures 22−3−22-b, which are taken from *A Guide for Control of Laser Hazards* (pp. 11−13), Cincinnati, OH: American Conference Governmental Industrial Hygienists, 1976.

B Microwave Equipment

Microwave equipment has become commonplace in homes (ovens) and in the work environment (test equipment). In addition, microwave equipment is used in a laboratory technique known as diathermy. Operators who perform this technique are exposed to microwave radiation. Microwave ovens also are heavily utilized in laboratories.

Facilities that use microwave equipment should follow the standards for exposure to microwave radiation developed by the Physical Agents Committee of ACGIH.

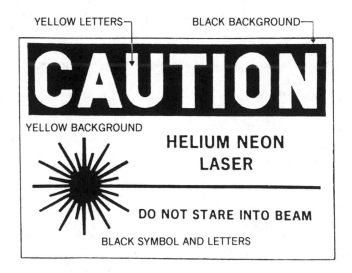

FIG. 22−3: Label for Low power lasers only.

FIG. 22−4: Label for Medium power visible laser having a total power output below 5 mW but a maximum beam irradiance of less than 2.5 mW/cm².

FIG. 22−5: Labels and signs for Medium- and high power lasers only (blank format).

FIG. 22−6: Medium and high-power laser only (with sample hazard control information filed in).

Microwave radiation affects molecular rotation and increases the kinetic energy of molecules in materials. The increase in kinetic energy has a thermal effect on the material which is of primary concern in body systems. Moderate heating of body tissues may cause the following: birth defects, testicular degeneration and partial or total sterility, cataracts, changes in immunological and endocrinal functions, and behavioral anomalies. In the medical application, microwave diathermy, the typical frequency used falls between 30 and 3000 MHz, and the primary body organ of concern is the lens of the eye. This is the critical wavelength region for eye cataracts.

All microwave installations should be maintained at a periodic frequency that is based on frequency of use and the manufacturer's recommendation. Devices used to measure microwave radiation work on the principle of converting microwave energy to heat and measuring heat changes with a sensitizing device. Two meters used in this type measurement are:

- Model 8100 Electromagnetic Radiation Survey Meter, built by Narda Microwave Corporation, Haupphuge, New York; and

- Microwave Survey Meter, built by Holaday Indusries, Hopkins, Minnesota.

All sampling of microwaves should be documented.

Microwave exposures are measured in the same units as ultraviolet radiation, milliWatts per square centimeter (mW/cm^2). The threshold limit values for microwave exposure are as follows:

- A power density of 10 mW/cm^2 limit is allowed for an exposure period of not more than 8 hours in one day.
- A worker may not be exposed to microwave energy that has a power density between 10 and 25 mW/cm^2 for more than 10 minutes in any one-hour period.
- Workers are not to be exposed to microwave energy that has a power density greater than 25 mW/cm^2.

The primary control for a microwave oven installation is a good periodic maintenance program in which the gaskets and general condition of the installation are both evaluated.

C Ultrasound

Ultrasound can be used in laboratories as an imaging technique. The ultrasound TLVs presented by the ACGIH are centered around the third-octave band at 20 kHz. Below 20 kHz the subjective effects of this noise are not present, and on the third-octave above 20 kHz hearing loss is possible because of the subharmonics that exist at these frequencies. These frequencies range from 10 to 50 kHz at 80−115 dB.

The ultrasound TLVs are presented in Table 22−2. If these TLVs are exceeded, control technologies utilized for overexposure to noise are applicable.

VI CHEMICAL HYGIENE PLAN CONSIDERATIONS

Sources of exposure to both ionizing and nonionizing radiation should be identified in the chemical hygiene plan. Practices for handling radioactive agents should be described in the hygiene plan. The following general requirements form the foundation of an appropriate program for most laboratories:

1. All weighing, handling and administration of radioactive aerosols, dusts, and gases are to be performed in facilities that have primary and

TABLE 22−2 Permissible Exposure Levels for
Ultrasound

Midfrequency of Third-Octave Band (kHz)	Sound Pressure Levels[a] One-Third Octave Band Level (dB re 20 μPa)
10	80
12.5	80
16	80
20	105
25	110
31.5	115
40	115
50	115

[a] Levels below which nearly all workers may be repeatedly exposed without adverse effect.

secondary containment, including glove box, and/or closed handling and process systems. Ventilation systems that remove contaminated air should have proper filtration in place. When a filter is replaced, it should be disposed as radioactive waste. The radiation safety officer should be consulted for proper protective equipment and procedures.

2. Preparation and handling of scintillation fluid are to be performed in a properly functioning hood.

3. Disposal of radioactive material is to be performed in accordance with municipal, state, and federal regulations.

Similarly, practices for control of nonionizing radiation should be described in the chemical hygiene plan.

RESOURCES

Martin and Harbison, *An Introduction to Radiation Protection* 2nd ed London: Chapman & Hall, 1979.

Shleien, Terpilak, *The Health Physics and Radiological Health Handbook*. Neicleon Lectern Associates, 1984.

U.S. Department of Commerce, National Bureau of Standards, *Safe Handling of Radioactive Materials*, Handbook, 92, 1964.

U.S. Nuclear Regulatory Commission, "Radiation Safety Surveys at Medical Institutions," *U.S. Nuclear Regulatory Commission Regulatory Guide 8.23*, Revision 1. Washington, D.C.: Jan., 1981.

CHAPTER 23

CONTROLLED SUBSTANCES

I INTRODUCTION

The federal government regulates all controlled substances. Thus, laboratories handling, storing, or testing such chemicals should become familiar all regulations enacted by federal, state, and local governments. This section highlights only controlled substance issues as dictated by the federal government.

Title 21 of the Code of Federal Regulations (CFR), Part 1300 to the end, contains the federal regulations that govern controlled substances, as enacted by the Federal Drug Enforcement Administration (DEA). Several official and quasi-official voluntary bodies are concerned with standards for the manufacturing, distribution, labeling, and advertising of drug products, including the Food and Drug Administration (FDA), the Federal Trade Commission (FTC), the Department of Justice (which is responsible for the administration of the Controlled Substances Act through the DEA), the Consumer Product Safety Commission, U.S. Pharmacopeial Convention, Inc., and the U.S. Adopted Names Council. All provide various services and functions in the management of controlled substances.

II SCHEDULE OF CONTROLLED SUBSTANCES

The federal government has divided the drugs that come under the jurisdiction of the Controlled Substances Act (Title II of the Comprehensive Drug Abuse Prevention and Control Act of 1970) into five schedules. This section presents some examples of drugs in each schedule. A more detailed listing is available in 21 CFR, Part 1308. The specific status of a drug is available from the regional office of the DEA.

Schedule I. Drugs in Schedule I are generally those that have no accepted medical use for treatments in the United States and also have a high potential for abuse. Some examples of such controlled substances are tetrahydrocannabinols, lysergic acid diethylamide (LSD), heroin, marijuana, peyote, mescaline, and benzylmorphine.

Substances listed in Schedule I are not available in prescription form. However, they may be obtained for research and instructional use or for chemical analysis. Application should be made to the Drug Enforcement Administration, Department of Justice (form 225). This application must be supported by a protocol that presents the proposed use of the controlled substance.

Schedule II. Schedule II drugs have a high potential for abuse, with severe liability to cause psychic or physical dependence. These drugs include, but are not limited to: opium, morphine, codeine, methadone, methylphenidate, and Dronabinol® (Marinol®). They also include certain other opioid drugs and drugs that contain amphetamines or methamphetamines as the single active ingredient or in combination with each other.

Other examples of narcotic drugs in Schedule II are Percodan (oxycodone), Pantopon, cocaine, Dilaudid (dihydromorphinone), Demerol (meperidine), Percobarb, and Percocet.

Some examples of Schedule II amphetamine drugs are Benzedrine, Dexedrine, Dexamyl, Eskatrol, and Biphetamine.

Some methamphetamine drugs in Schedule II are Methedrine, Desoxyn, and Ambar.

Some depressants are Amytal (amobarbital), Seconal (secobarbital), Tuinal (amobarbital and secobarbital), and Nembutal (pentobarbital).

Schedule III. Schedule III drugs include those with a potential for abuse, but less than that of either Schedule I or II drugs. Abuse of these drugs may lead to moderate or low physical dependence or high psychological dependence. Schedule III drugs includes compounds that contain limited amounts of certain opioid drugs as well as certain non opioid drugs. Opioid-containing compounds include Empirin® compound with codeine, Phenaphen® with codeine, Soma® with codeine, Codempiral #2®, Donnagesic paregoric, and Tussionex®.

Non narcotic drugs in Schedule III include Noludar® (methyprylon), Doriden® (glutethimide), Butisol® (butabarbital), Fiorinal®, nalorphine, and certain barbiturates (except those listed in another schedule).

Schedule IV. Schedule IV drugs have a low potential for abuse. They may lead only to a limited physical or psychological dependence when compared with Schedule III compounds.

Examples of drugs in Schedule IV include barbital, phenobarbital, and chloral hydrate (Noctec®, Felsules®, Kessodrate®, Somnos®), also, paraldehyde, meprobamate (Equanil®, Miltown®), and Placidyl® (ethchlorvynol), Librium®, Valium®, Darvon®, and Talwin®.

Schedule V. Schedule V drugs have a potential for abuse less than those listed in Schedule IV. When prepared, Schedule V drugs have moderate quantities of certain narcotic drugs added for use, for example, in treating coughs or diarrhea. These may be distributed without a prescription order.

Schedule V drugs include preparations formerly known as "exempt narcotics," such as cough syrups that contain codeines. Examples include Robitussin-AC® and terpin hydrate with codeine.

The federal government is responsible for overseeing any changes in thcsc schedules. The government can effect changes by additions or deletions, or by upgrading or downgrading any of the controlled substances.

III IMPLEMENTING CONTROLLED SUBSTANCE PROGRAMS

The federal regulations outlined in this section appear in full in Title 21 of the Code of Federal Regulations (Part 1300−end). All sections that pertain to laboratory use of controlled substances are briefly discussed in the following subsections.

A Registration of Manufacturers, Distributors, and Dispensers of Controlled Substances

1 Requirements for Registration

Every person who manufactures, distributes, or dispenses any controlled substance is required to obtain a registration every year, unless exempted by law, or when parts 1301.24−1301.29 of Title 21 apply. A separate registration is required for any laboratory or individual that conducts research with controlled substances listed in Schedules II−V. This regulation also applies to all laboratories or individuals that conduct research and instructional activities with the substances in Schedule I.

Before a laboratory is permitted to conduct research with one or more of the controlled substances listed in Schedule I, it must submit form 225 (available from the DEA) with three copies of a research protocol. The laboratory must prepare the protocols in the format described in Section .33 of 21 CFR 1301 (21 CFR 1301.33). The applications for registration may be submitted at any time.

Before using the compounds listed in Schedules II–V, a laboratory should also submit form 225 to the DEA.

Even a person who already is registered with the DEA must reapply for registration annually on form 227. This form is also available from the DEA.

2 Security Requirements

All registrants and applicants for registration must provide effective controls and procedures to guard against the diversion and/or theft of all controlled substances. The security requirements are set forth in 21 CFR 1301.72–1301.76. These are standards for the physical security controls and operating procedures needed to prevent theft.

Practitioners should store controlled substances listed in Schedules I–V in a securely locked, substantially constructed cabinet or safe. However, pharmacies and institutional practitioners may dispense controlled substances listed in Schedules II–V throughout their stock of non-controlled substances in a way that obstructs theft or diversion. This also applies to nonpractitioners authorized to conduct research or perform chemical analysis under separate registration. All controlled substances must be stored in a secure area where theft or diversion can be prevented.

Nonpractitioners should store controlled substances listed in Schedules I and II in a safe steel cabinet or a vault that meets specific requirements prohibiting their theft or diversion. They should store controlled substances listed in Schedules III–V in a safe steel cabinet, vault, storage building, cage, or other approved enclosure that meets specific requirements prohibiting theft or diversion (see Sections 71 and 72 of 21 CFR 1301). Also, nonpractitioners should screen their employees to try to prevent a drug security breach. These regulations and procedures are needed as a business necessity to ensure the overall security of controlled substances. Examples of typical questions asked for security purposes are found in 21 CFR 1301.90.

A laboratory should see that the amount of any controlled substance outside the secured storage area does not exceed its daily needs. To minimize the possibility of internal diversion, a laboratory should limit access to storage areas for controlled substances to a minimum number of employees. Also, a laboratory must notify a regional office of the DEA of any theft of any controlled substance, or of significant loss.

3 Labeling and Packaging Requirements for Controlled Substances

All commercial containers of controlled substances must be labeled with a symbol that indicates the schedule of the drug(s). The following symbols designate the schedule:

Schedule I	CI or C-I
Schedule II	CII or C-II
Schedule III	CIII or C-III
Schedule IV	CIV or C-IV
Schedule V	CV or C-V

The word "schedule" need not appear on the label. Also, no distinction between narcotic and nonnarcotic substances is necessary. The "C" symbol is not required on the surface of a carton or wrapper if the symbol is legible through the carton or wrapper, if the container is too small and the symbol is printed on the box or package from which the container was removed, or if the controlled substance is being utilized in clinical research (e.g., blind or double-blind studies).

The symbol does have to be displayed prominently on the label of a commercial container, and it must be at least twice as large as the largest type. Other specifications for the location and size of the symbol may be found in 21 CFR 1302.04−1302.05. The most important part of labeling is that the symbol be clear and large enough for easy identification without removing the container from its storage area.

For controlled substances listed in Schedule I or II, and for any narcotic controlled substance in Schedule III or IV, a wrapper or seal must be securely affixed to the stopper, cap, lid, covering, or wrapper. Tampering or opening of the container is indicated if the wrapper or seal is not intact.

4 Aggregate Production and Procurement Quotas and Individual Manufacturing Quotas

The quota requirement mandates that the DEA annually estimate the total quantity of Schedule I or II drugs that must be manufactured during a given calendar year to provide sufficient quantities for estimated medical, scientific, research, and industrial needs. Various factors are involved in the estimate, and every year the Federal Register publishes its estimates for procurement quotas. Several categories of persons, including laboratory personnel, unless they are involved in the manufacture of these compounds, need not apply for a procurement quota; see 21 CFR 1303.12.

B Records and Reports of Registrants

1 Inventory Requirements

All persons who are registered must keep records and file reports on all controlled substances. For example, a researcher who manufactures a controlled substance must keep the following records: quantity manufactured, date of distribution, quantity distributed, and transfer documentation. All records must be maintained in one consolidated record system.

A researcher or laboratory must store all controlled substances in one place. Also, every 2 years it is necessary to conduct an inventory, regardless of whether the substances were manufactured by, imported by, or distributed to other researchers, or destroyed by chemical analysis.

However, a registered person using any controlled substance in research conducted in conformity with an exemption granted under Section 505(i) or 512(i) of the federal Food, Drug, and Cosmetic Act [21 USC 355(i) or 360(j)] at a registered establishment that maintains records in accordance with either of those sections is an exception. Such persons are not required to keep records if they notify the DEA of the name, address, and registration number of the establishment maintaining such records.

Each inventory for the laboratory must contain a complete and accurate record of all controlled substances on hand on the date the inventory is taken. A separate inventory must be made by a registrant for each registered location. Also, a registrant must perform a separate inventory for each independent activity for which he or she is registered.

A registrant may conduct an inventory on a date that falls within 4 days of his biennial inventory date if he notifies the proper authority. A registrant must document an inventory in a written, typewritten, or printed form. The registrant must take an inventory of all stock of controlled substances on hand every 2 years following the date of the initial inventory.

2 Special Requirements for Inventories of Dispensers and Researchers

For each controlled substance in finished form, the following information must be included in the inventories:

name of substance,

finished form of substance,

number of units or volume of each finished form in each commercial
container, and

number of commercial containers.

The following information is necessary with respect to substances that are
found to be damaged, defective, or impure during the inventory:

name of substance,

total quantity of substance, and

reason for substance being maintained.

For Schedule I and II substances, the inventory must include an exact
count or measure. For controlled substances in Schedules III—V, the
inventory must make an estimated count or measure of the contents of
each container. Unless the container holds more than 1000 tablets or
capsules, the inventory must disclose its exact contents.

3 Inventories of Chemical Analysts

Special requirements and information are needed on the inventories of
registrants who perform chemical analysis. For further information on
these inventories, see 21 CFR 1304.19.

4 Continuing Records

Every registrant is required to maintain, on a current basis, a complete
and accurate record of each controlled substance manufactured, imported
exported, received, sold, delivered, or otherwise disposed of. However,
no perpetual inventory is required.

5 Reports

Continuing records for dispensers and researchers should include the
following information:

name of substance,

number of units or volume of finished form,

number of commercial containers,

number of units or volume dispensed, and

number of units or volume destroyed.

This information should be updated regularly and kept in one area for easy identification and availability.

C Order Forms

The federal government requires reports on controlled substances from various manufacturers. The format and content of the reports are discussed fully in 21 CFR 1304.38−1304.41. Special reports are needed from the following:

manufacturers who import opium,

manufacturers who import medicinal coca leaves,

manufacturers who import special coca leaves,

manufacturers of bulk materials or dosage units,

packagers and labelers.

distributors,

manufacturers who import poppy straw or concentrates of poppy straw.

Order forms (form 222c) available from the DEA are required for the distribution of any controlled substance listed in Schedule I or II. For information about who is required to submit filled-out forms, how to obtain order forms, procedures for filling and endorsing order forms, reporting stolen forms, and other details, see 21 CFR 1305.04−1305.16.

D Prescriptions

In a laboratory setting, for research purposes prescriptions would not be of concern unless a clinical trial were being performed. For further information on prescriptions, refer to 21 CFR 1306.

E Miscellaneous

Special exceptions for the manufacture and distribution of controlled substances are discussed in 21 CFR 1307.11−1307.15. For the most part, such exceptions would not apply to laboratories using controlled substances.

F Disposal of Controlled Substances (21 CFR 1307.21−1307.31)

Under no circumstances can any controlled substance in Schedules I−V be destroyed without approval of DEA personnel.

Any registrant who desires to dispose of any controlled substance must notify, in writing, the special agent in charge of administration in the area. The registrant must keep proper records and file three copies of each notification on the proper forms.

Non registrants must submit, in writing, certain specified information, including the name and address, and the name and quantity of the controlled substance to be disposed of, to the DEA. Then the special agent in charge of administration in the area shall authorize and instruct the applicant in the proper controlled substance disposal procedures.

IV SCHEDULES OF CONTROLLED SUBSTANCES — SPECIAL EXEMPTIONS

Schedules I–V are described in detail in Part 1308 of 21 CFR. Certain chemical preparations are exempt from the application of all or any part of the information above if the preparation or mixture is intended for laboratory, industrial, or educational, or special research purposes and not for general administration to a human being or animal. Under such exemption, the controlled substance also should be packaged in a way that presents no significant abuse potential. Filers for exemption must notify the DEA and supply the agency with proper information on the controlled substance.

V PIPERIDINE REPORTING AND PURCHASER IDENTIFICATION

Any person who distributes, transfers, sells, ships, or imports piperidine, its salts, and/or acetyl derivatives must maintain proper records and identification. Records of all piperidine purchases and distributions must be maintained (21 CFR 1310).

VI REGISTRATION OF IMPORTERS AND EXPORTERS OF CONTROLLED SUBSTANCES

Specialized information for importers and exporters of controlled substances can be found in Parts 1311 and 1312 of 21 CFR. For the most part, this would not apply to laboratories.

VII ADMINISTRATIVE FUNCTIONS, PRACTICES, AND PROCEDURES

The DEA may inspect at any time, the facilities and/or records of locations where controlled substances are handled. Part 1316 of 21 CFR discusses protection of researchers and research subjects on grounds of confidentiality and protection from the Controlled Substance Act.

Further information on implementation of a controlled substances program is available by referring to Title 21 CFR (Part 1300–end), or, by writing to:

Registration Unit
Drug Enforcement Administration
Department of Justice
Post Office Box 28083
Central Station
Washington, DC 20005

VIII CHEMICAL HYGIENE PLAN CONSIDERATIONS

The chemical hygiene plan should describe the precautions to be taken for storage and handling of controlled substances. In particular, requirements for storage and security, inventories, and recordkeeping should be described.

RESOURCES

American Medical Association, *Drug Evaluations*, 6th ed. Chicago: American Medical Association, 1986.

Code of Federal Regulations 21, Office of the Federal Register, National Archives and Records Service, General Services Administration, 1986.

Gilman, A. G., L. S., Goodman, T. W., Rall, and F. Murad, *Goodman and Gilman's The Pharmacological Basis of Therapeutics*, 7th ed. New York: Macmillan, 1985.

Klaasen, D. C. *et al.*, *Casarett and Doull's The Basic Science of Poisons*, 3rd ed. New York: Macmillan, 1986.

Mathieu, M. P. (ed.), *New Drug Development: A Regulatory Overview*. Washington, DC: OMEC International, 1987.

O'Reilly, J. T. *Regulatory Manual Series: Food and Drug Administration*. New York: McGraw-Hill, 1985.

Osol, A. *Remington's Pharmaceutical Sciences*, 16th ed. Easton, PA: Mack, 1980.

CHAPTER 24

WASTE MANAGEMENT

I INTRODUCTION

Laboratory management is responsible for the development of a waste management program that will ensure the safe handling and disposal of all laboratory wastes. This program should reflect the specific activities of the individual laboratory and should incorporate all applicable federal, state, and local regulations.

An effective program for managing laboratory wastes has two basic goals: (1) to operate the laboratory in compliance with all applicable regulations and (2) to manage the wastes generated in a manner that protects employees, the citizenry, and the environment. A laboratory should communicate these goals through policy statements and detailed standard operating procedures (SOPs), which form the basis for laboratory personnel training. The laboratory should formalize these policies and procedures to the extent practical and incorporate them into its operating manual. The laboratory can use this manual as part of its training program, and employees already trained can use it as a reference.

The laboratory must update its waste management program periodically, so that it reflects changes in laboratory activities and governmental requirements, as well as current best management practices. Laboratory personnel should take refresher courses periodically to keep themselves current with the latest waste management policies and procedures. The laboratory should review these policies and procedures from time to time with its employees. The object of such reviews would be to solicit employee feedback and ensure a consistent understanding of the laboratory's waste management program.

II IMPLEMENTATION OF A WASTE MANAGEMENT PROGRAM

The waste management program should reflect the specific activities of the laboratory, clearly stating the laboratory's objectives with respect to waste management. Generally, these program objectives should include:

- protection of the health and safety of the workers;
- protection of the environment;
- protection of the health and safety of the public;
- compliance with federal, state, and local regulations; and
- adherence to sound management practices.

The laboratory should communicate its goal of protecting the health and safety of workers and the public, along with protecting the environment, through a set of program policy statements and its obvious day-to-day commitment to them. Standard operating procedures deal with the actual steps necessary to carry out these goals, achieve regulatory compliance, and minimize wastes. Both the policies and the SOPs are integral parts of the laboratory's operating manual and training program.

A Regulatory Compliance

Regulations that govern the handling of waste streams are derived from federal, state, and local authorities and control chemical release to the air, water and land. The impacts on the environment from the failure to control hazardous wastes are covered by federal legislation that includes the following:

- Clean Air Act,
- Clean Water Act,
- Toxic Substances Control Act (TSCA),
- Resource Conservation and Recovery Act (RCRA) and
- Superfund, the Comprehensive Environmental Response, Compensation and Liability Act (CERCLA).

The primary set of regulations that pertain to chemical wastes is known as the Resource Conservation and Recovery Act (RCRA). These regulations, which form the basis for most waste management programs, are discussed in detail below.

The Clean Air Act requires the nation's states to develop their own plans for complying with federal air guidelines. Each state, in turn, is to develop a permit system for air pollution sources. Operations at laboratories that may require air pollution permits include incineration of animal carcasses and other wastes (possibly releasing radioactive materials, biological hazards, hydrocarbons, or particulates), vents from hoods (volatile organics, corrosive gases), and possibly industrial boilers (particulates, hydrocarbons). Requirements vary from state to state. Thus a laboratory should contact the appropriate environmental agency to determine what operations require air permits. If needed, the permit may require the periodic measurement of emissions and the maintenance of air pollution control equipment. These required actions should be reflected in the laboratory's standard operating procedures.

The Clean Water Act is a federal program that empowers the states to issue water discharge permits. If a laboratory discharges to a surface water body, such as a river, lake, pond, or stream, it may require a National Pollutant Discharge Elimination System (NPDES) permit. This permit, which is usually administered by the state, may set limits for certain pollutants. It may also specify monitoring and reporting requirements and water pollution control equipment maintenance.

If a laboratory discharges its waste to a sewer that is connected to treatment works, or a publicly owned treatment works (POTW), the POTW is responsible for meeting the conditions of its NPDES permit. To achieve this goal, the POTW may require a permit or specify limitations for all those it serves. A facility that uses a septic system may also discharge wastes that could impact water quality. Some states require a permit for all systems that may impact groundwater quality, including septic systems; such facilities may be required to obtain what is often called a discharge-to-groundwater permit.

In all cases, the regulatory concern centers on any organic or inorganic (especially heavy metal) wastes that may enter the wastewater system from the contributing facility. Other criteria, such as suspended solids, pH, fecal coliform, and oil/grease content of the aqueous wastes, may also apply. A facility unsure of the quality of its discharge should contact the state environmental agency to determine its need for a water permit. As with air permits, any requirements stated in the water discharge permits should be incorporated into the laboratory's SOPs.

The Toxic Substances Control Act has only limited applicability for most laboratories. One set of requirements that may apply concerns the use of polychlorinated biphenyls (PCBs) on site. Before the late 1970s, manufacturers often used PCBs as a dielectric material in electrical equipment, and PCBs in this form may be present in older transformers, capacitors, or switching gear located on site. Even if a laboratory does not own or operate such equipment (responsibilities that typically belong to the electric utility), the laboratory may be responsible for determining whether the owner or operator is operating any PCB equipment. This obligation may mean that the laboratory has to establish a special monitoring and maintenance program. A laboratory should be able to document what electrical equipment it uses, regardless of whether it is being properly managed by the owner/operator.

RCRA covers the management of regulated hazardous wastes and specifies several basic responsibilities for generators of wastes. According to the act, a waste generator must determine whether the wastes are regulated as hazardous, determine its generator status by calculating the amount of hazardous waste generated, and comply with the management regulations according to the appropriate generator status. RCRA gives the generator "cradle to grave" responsibility for the wastes it produces. In fact, the act makes the generator liable for its waste materials even after they have reached their final disposition.

A waste is any solid, liquid, or gaseous material that is no longer used and will either be recycled, disposed of, or stored in anticipation of treatment or disposal. Part 261 of Title 40 of the Code of Federal Regulations (40 CFR 261) defines hazardous wastes in two basic parts: a waste is regulated as hazardous (1) if it is specifically listed by name or by category, or (2) if it meets one of four characteristics namely, corrosivity, ignitability, reactivity, and extraction procedure (EP) toxicity. Typical laboratory wastes that are regulated as hazardous include acids and bases, heavy metals and inorganic materials, ignitable wastes, reactives (oxidants), and solvents.

Some wastes are considered to be "acutely hazardous" and even small amounts are regulated in the same way as large amounts of other wastes because they are thought to be so dangerous. These wastes, which include certain pesticides and dioxin-containing wastes, are also specified in 40 CFR 261.

The first basic responsibility of any business is to find out whether it generates hazardous wastes. It can make this determination by identifying all waste streams and applying the criteria for hazardous waste against each waste stream. If a business determines that it is generating a hazardous waste, it must quantify the amount to ascertain the regulatory status that applies to the location. If the location generates more than 1000 kg of hazardous waste (or 1 kg of acutely hazardous waste) per month, the location must meet all the requirements of a generator a specified in 40 CFR 262 and as modified by each state. If the location generates less than 1000 kg but more than 100 kg of hazardous waste (and less than 1 kg of acutely hazardous waste), only some of the generator requirements may apply. If the location generates less than 100 kg of waste in any calendar month, it is exempt from the formal management program requirements, as long as it accumulates no more than 1000 kg of waste on site. Note that some states have reduced these quantity criteria.

All generator categories are required to manage their wastes safely, and they must send them to approved handling facilities. Generators of more than 100 kg of waste per month must obtain an identification number through the state environmental agency or from the regional office of the federal Environmental Protection Agency (EPA). When sending wastes off site, generators must pack them according to Department of Transportation (DOT) specifications, as noted in 40 CFR 262 and 49 CFR 172.

Laboratories may choose to pack their own wastes, or they may have the hazardous waste contractor do this, as part of the removal service. The other requirements for regulated generators deal with the accumulation and storage of the wastes, the tracking and reporting of these wastes, and the actions to be taken in an emergency.

The laboratory's SOPs should reflect all requirements, such as labeling, storage container specifications, management and inspection of storage areas, accumulation times, manifest system, and reporting. A location may need a permit to treat, dispose of, or store hazardous wastes if these activities are beyond those allowed for generators. In such cases, the location should contact the state environmental agency to gain a complete understanding of all applicable requirements.

Another set of environmental regulations that apply to facilities that manage hazardous waste was established by the comprehensive act called CERCLA, or Superfund. While the majority of these regulations deal with waste disposal sites, a portion of them apply to the release of regulated hazardous wastes to the environment. 40 CFR 302 explains the required reporting steps and defines the minimum reportable release (reportable quantities, or RQs) in the event that a regulated hazardous waste is released. This notification system should be included in the laboratory's SOPs and may even apply to nonwaste materials the laboratory handles.

The laboratory is also responsible for determining whether any state or local regulations apply to its waste activities. For example, California recently adopted a set of requirements (known as Proposition 65) that severely limit the discharge of many pollutants. Some communities also have special reporting or handling requirements that may affect a laboratory's operations. The state environmental agency should be able to identify state and local requirements, or provide a local agency contact that can define applicable requirements. These provisions should also be included in the SOPs. A listing of state environmental agencies is appended to this chapter.

B Best Management Practices

Beyond compliance, a set of generally accepted waste material management practices exists, which makes compliance easier to achieve, reduces waste-related costs, and minimizes adverse environmental impact. These practices relate to the quality and thoroughness of the waste management systems and to principles of "good housekeeping" and efficient material use. Some of these practices may be required to some extent by the regulations, but most are not specifically required. The success of these practices depends on the commitment of a laboratory's management.

Required waste management systems may be improved beyond achieving compliance through clear documentation and internal communication of the waste management systems and individual responsibilities. Additional improvement may be obtained by designing systems that fit the particular organization and style of a given laboratory, and by recognizing the contributions of those who do the actual work. The clear communication of a system's approaches and responsibilities is critical and should be made a part of both the manual and training programs of a laboratory. Management commitment to high quality programs can make the difference between a program that exists on paper and meets only the minimum requirements, and a program that is effective and actively manages waste materials. Opportunities to enhance required programs include: communication of regulatory requirements (training), labeling, management of storage areas and recordkeeping.

When a facility generates wastes, it can minimize the volume and cost of these wastes by following a few basic "housekeeping" principles:

- Do not mix nonhazardous and hazardous wastes. Such a mixture will have to be regulated as a hazardous waste, thus increasing the cost and responsibility associated with the waste.
- Segregate hazardous wastes. This practice may be required to prevent the mixing of incompatible wastes, but it also makes sense to preserve the waste properties, and thus make recycling or treatment easier and less expensive. For example, do not mix halogenated with non-halogenated solvents.

- Avoid spills or leaks. Through a preventive maintenance program and inspections, a facility can minimize spills and leaks, thus reducing the amount of wastes generated. (Spill clean up materials for hazardous wastes are regulated as hazardous wastes.)

In a laboratory environment, management can implement these principles by providing separate, clearly identified containers for nonhazardous solids, hazardous solids (including contaminated disposable protective clothing), nonhazardous liquids, hazardous halogenated liquids, and hazardous nonhalogenated liquids. (Of course, only compatible materials should be placed in the same container.) This practice can help avoid creating "orphan" wastes (i.e., wastes that are of unknown origin and characteristics and are therefore difficult to handle). The facility should handle any "orphan" material as a hazardous waste and should characterize it to the degree possible. In addition, the facility should examine the storage of chemicals, as well as experimental apparatus, to identify potential spills and leaks and to plan how to avoid them.

Certain waste management approaches complement these practices. The first is the communication and implementation of a waste management hierarchy that specifies that waste handling options are to be accorded the following priorities: recycling, reuse, recovery, treatment, and disposal. This approach reduces the impact on the environment by favoring methods that keep the materials in a system and precludes their release to the environment. Another approach to maintaining control over wastes and their associated liability is to screen and monitor the facilities that recycle, treat, or dispose of the wastes generated at a given location. This type of program gives the user some assurance that wastes are being appropriately handled and reduces the threat of possible environmental liabilities.

In many laboratories, biological wastes require special consideration. The laboratory must manage these wastes in such a way that the presence of any viable organisms, carcinogens, and/or radioactive elements can be accounted for. These wastes are discussed in Sections III and IV.

Of course, the best way to manage wastes is not to generate any in the first place. This approach is the basic concept behind waste minimization. A laboratory can minimize wastes by substituting nonhazardous chemicals for hazardous ones, carefully controlling chemical inventories, and using chemicals and recovery systems efficiently. Each laboratory should strive to minimize its wastes to the extent practical and should review its waste minimization approaches periodically to incorporate new laboratory operations and new waste minimization techniques.

Laboratories have a number of opportunities to replace the methods and chemicals they are currently using with less toxic materials and methods. Each laboratory should review its hazardous and/or toxic raw materials to determine the possibility of substitution with less hazardous or toxic materials. For example, a laboratory can often reduce the hazards associated with glassware cleaning and solvent extractions. If a laboratory uses a nonhazardous solvent in the place of a solvent that would result in a hazardous waste, it can reduce the amount of hazardous waste it generates. Most laboratories have replaced benzene with toluene as a solvent to reduce health and environmental impacts.

Laboratories can manage their inventories to minimize the generation of wastes from the disposal of expired, out-of-specification, or otherwise degraded materials by adopting a "first in/first out" policy (i.e., using older chemicals before newer ones). A part of this policy is to clearly mark the expiration dates on the containers, along with the date the material was received, and to conduct periodic inventories of materials. (Such inventories afford a convenient time to clear out all hazardous wastes from the laboratory.) This information can serve to help the procurement process by documenting the usage rate, and costs can be reduced by eliminating unnecessary chemical purchases. Not only are expired chemicals unusable and a waste of resources, but some chemicals form dangerous degradation products as well. Thus, each laboratory should have an inventory policy and a system to implement it.

Using chemicals efficiently is a goal every laboratory should have to reduce reagent costs as well as waste disposal costs. Such a policy must be implemented on a case-by-case basis, relying on the judgment of the investigators and the needs of the study. Some possible methods are to reduce the amount of solvent during glassware cleaning and to implement higher efficiency extraction procedures.

Recovering materials is another approach that can minimize wastes. A laboratory can regenerate solvents through distillation, or otherwise removing contaminants. Also, it can use such techniques as ion exchange, ultrafiltration, reverse osmosis, centrifuging, and distillation to reduce both raw material usage and waste disposal costs. Each laboratory should also review the materials it uses to identify opportunities to replace chemicals with substitutes that are more readily recoverable. Such approaches may result in a net savings to the laboratory—an approach that simply makes good economic sense.

A final option for waste management may be on-site treatment, such as pH adjustment or precipitation. These options might allow a laboratory to include all or part of a given waste stream in its wastewater flow. This would reduce the volume of wastes regulated as hazardous. In some cases, however, on-site treatment is regulated under RCRA; or it may be unacceptable to the local water or sewer authority. Even if treatment is allowed, such approaches may not be considered to be best management practices if they increase the quantity of pollutants in the environment.

The laboratory should contact the state environmental agency to ascertain whether any on-site treatment is acceptable.

Several references are available that will help laboratory employees to handle waste materials safely. The first source of information is the material safety data sheet (MSDS) that accompanies the raw materials at the time of delivery. The MSDS specifies the properties and hazards of the chemical and also prescribes the manner in which a laboratory should handle the resultant waste. Both quality and extent of the information on MSDSs vary from supplier to supplier, and thus these documents may have to be supplemented. Additional reference materials can be obtained from organizations like the National Fire Prevention Association (NFPA) (e.g., *Hazardous Chemical Data* and *Manual of Hazardous Chemical Reactions*) and from trade and professional organizations.

III MANAGEMENT OF RADIOACTIVE WASTE

A laboratory should incorporate the management of radioactive wastes into all aspects of its hazardous waste management program. It should also incorporate the topic of radioactive waste into its volume waste reduction policies and involve individuals who are knowledgeable about handling radioactive materials. However, the actual handling of such waste is an issue separate from hazardous waste.

Radioactive waste results when a laboratory uses radioisotopes, usually as tracers. A radioactive material is one that contains at least 0.005 microcurie (μCi) per gram of material, or per milliliter if a liquid. Below this limit, a laboratory may discard materials without regard to their radioactive component; above this limit disposal becomes a strictly regulated issue.

Before disposing of any radioactive waste, a laboratory should contact a waste broker. Waste brokers accept packed waste drums and transport them to a final disposal site. Currently, there are only three such sites authorized for the disposal of radioactive waste: Richland, Washington; Barnwell, South Carolina; and Beatty, Nevada.

Regulations regarding waste packaging and disposal of radioactive material can be found in 10 CFR 20 State Disposal Site Regulations and 49 CFR 173. Nuclear Regulatory Commission (NRC) regulations (10 CFR 20) include the disposal of radioactive materials down a drain and alternative methods. The state in which the disposal site resides sets the allowable quantity for radionuclides in drums, and the DOT regulations (49 CFR, Part 173) describe the packaging and labeling requirements for waste drums being transported for disposal.

A waste broker who handles radioactive waste will tell the laboratory how to properly segregate, package, and label the disposal containers. These requirements may differ for various states, depending on the waste site to which the broker delivers.

Each site requires the waste generator to apply for a permit from the state to which its waste is to be delivered. Waste brokers ask a facility for its site user's permit before accepting its waste. The permit number issued by the state will appear on all pamphlets.

As with hazardous waste, the *generator* is ultimately responsible for the waste, even if it is handled by a contractor. Therefore, it is very important that the laboratory make sure that its contractor handles its waste responsibly and legally.

A Radioactive Waste Packaging

When preparing radioactive waste for disposal, a laboratory should pack the waste into DOT-approved 55-gal steel drums by one of five categories: dry solid waste, absorbed liquids, animal carcasses, aqueous vials, and liquid scintillation vials.

A laboratory should place all such waste into a thick plastic bag before putting it into the drum. Absorber material is used to line the bottom between the drum and the plastic bag. Typical absorbents include Speedi-Dry® and diatomaceous earth. Regulations do not allow the mixing of waste types within a single drum.

The five waste types are packaged as follows:

- Dry solid waste is placed inside a thick plastic bag, along with a layer of absorbent in the bottom of the drum.
- Absorbed liquids require placement of a layer of absorbent in the bottom of the drum before the bag is installed. With the plastic bag lining in the drum, the liquids are poured into alternating layers of 12-in.-thick absorbent. The drum must contain enough absorbent to absorb at least twice the volume of radioactive liquid contents.
- Animal carcasses are placed inside 30-gal, plastic-lined drums with lime and absorbent in a 1 10 ratio. The 30-gal drum is then placed in a 55-gal drum with enough absorbent to fill all the space between drums. Both drums are sealed.
- Aqueous vials are placed unopened in either 30- or 55-gal plastic-lined drums with 3 in. of absorbent placed in the bottom of the drum before the liner is installed. The vials (not to exceed 50 cc/vial) and absorbent are placed in alternating layers not exceeding 6 in. in depth. The top layer must consist of at least 3 in. of absorbent material.

The procedure for packaging liquid scintillation fluids is described next.

B Liquid Scintillation Fluid Packaging

Laboratories frequently use liquid scintillation to identify concentrations of radiolabeled compounds. The most frequently used radioisotopes are carbon-14 and tritium. Presently waste brokers are accepting only scintillation wastes containing certain types of radioisotope.

Industry designates liquid scintillation fluids to be hazardous materials (flammable liquids), radioactive materials, or both, for transportation considerations. A material's radioactivity is determined on an activity-per-unit-weight-of-fluid basis. Also decided by the presence or absence of radioactivity are the specifications for packaging the fluids. Table 24−1 gives the limits and classifications.

Scintillation wastes are incinerated. All drums require the addition of absorbent material before the liner can be placed in the drum. Disposers usually put unopened vials of fluid in a plastic bag before placing them in a lined drum, thus creating a double plastic lining. Then, before transportation, they seal the drum.

C Recordkeeping

The Nuclear Regulatory Commission requires laboratories to keep detailed records of all materials they dispose of — whether down a drain or into a drum. The U.S. Department of Transportation requires that the manifest accompanying the waste give a detailed description of the type of waste and the amount of radioactive material in each drum shipped to a disposal site.

TABLE 24−1 Classifications of Liquid Scintillation Wastes

Isotope	Quantity	Classification	Drum Specification
All isotopes (except ^{14}C and ^{3}H)	<0.002 μCi/g (except ^{14}C and ^{3}H)	Flammable liquid	Strong-tight: 49 CFR 173.411
	>0.002 μCi/g	Radioactive material	Type A: 49 CFR 173.411−415
^{14}C and ^{3}H	<0.05 μCi-g	Exempt (refer to RCRA)	Strong-tight

IV BIOHAZARDOUS WASTE MANAGEMENT

As in the cases of chemical and radioactive wastes, laboratories must handle biohazardous wastes independently. However, unlike other wastes, biohazardous wastes must be decontaminated before disposal, and this is the laboratory's responsibility. The point at which biohazardous agents are to be decontaminated depends on the biosafety level. However, the type of decontamination depends partly on the actual agent and partly on personal preference. Any decontamination method should include the following general procedures:

- Biohazardous materials should be sterilized before regular washing or disposal.
- A strong oxidizing material should never be autoclaved with paper, cloth, or other organic materials, because an explosion may be engendered.
- Floors and laboratory surfaces should be disinfected regularly.
- Floors should not be swept.
- Decontamination procedures should be assessed for compatibility with materials that come in contact with disinfectant (e.g., gloves, bench tops, plastics, and floor tiles).

Specific decontamination methods are described below.

1. *Wet heat.* Steam sterilization in an autoclave at a pressure of approximately 15 psi and a temperature of 121°C (205°F) for at least 15 minutes. Autoclaves should be calibrated for temperature and pressure, and monitored with a biological indicator, such as *Bacillus stearothermophilus* spores. It is important that the steam and the heat be made to contact the biological agent. Therefore, bottles containing a liquid material should have loosened caps, or cotton plug caps, to allow for steam and heat exchange within the bottle.

2. *Dry heat.* This form of sterilization generally requires temperatures of 160–170°C (320–338°F) for 2–4 hours. Again, it is important that the items be arranged in the autoclave with regard to heat transfer.

3. *Liquid disinfectants*

- Alcohol. Ethanol or isopropanol (70–85%) can effectively denature proteins, but not lipids. Therefore, when using alcohol, it is important to know whether the agent is composed mostly of lipid.
- Chlorine. A 1:100 dilution of bleach is a very effective disinfectant

against all microorganisms. This disinfectant may be effective against several life-threatening viruses, including the AIDS virus. It is important to remember, however, that this compound will lose its effectiveness over time and that even at a 1:100 dilution, it is corrosive to metals and even stainless steel.

• Iodine. Wescodyne is an iodine-based disinfectant often encountered in laboratory. Dilutions of 3 oz. in 5 gal of water are recommended for general laboratory cleanup, and a 1:10 dilution in 50% ethanol is recommended for hand washing.

• Formaldehyde. Formaldehyde is an effective disinfectant at 5% active ingredient concentration. Vapors of formaldehyde solutions may be irritating and should not be inhaled. Formaldehyde solutions should be made fresh to ensure effectiveness.

• Phenolic compounds. These disinfectants are not generally effective against bacteria, but they are usually used as disinfectants against rickettsia, fungi, and some vegetative bacteria. Phenol alone is not a good disinfectant because of its physical properties.

4. *Gases/Vapors*

• Formaldehyde. One method of decontaminating a biological safety cabinet involves heating paraformaldehyde inside a sealed cabinet. This creates a vapor that can travel throughout the cabinetry. Users should consult the cabinet manufacturer, or manual, for details of decontamination. A similar method may be employed in the decontamination of carbon dioxide incubators and even laboratories. Such procedures require considerable caution; use of gloves, masks, and coats is recommended.

V CHEMICAL HYGIENE PLAN CONSIDERATIONS

The laboratory's policies and practices for disposal of hazardous substances should be described in the hygiene plan. The plan should address definition of hazardous waste, waste collection, waste storage, labeling, recordkeeping, and disposition of waste.

ADDENDUM

STATE ENVIRONMENTAL AGENCIES

ALABAMA

Alabama Department of
  Environmental Management
Land Division
1751 Federal Drive
Montgomery, Alabama 36130
(205) 271−7730

ALASKA

Department of Environmental
  Conservation
P.O. Box 0
Juneau, Alaska 99811
Program Manager: (907) 465−2666
Northern Regional Office
  (Fairbanks): (907) 452−1714
South-Central Regional Office
  (Anchorage): (907) 274−2533
Southeast Regional Office
  (Juneau): (907) 789−3151

AMERICAN SAMOA

Environmental Quality
  Commission
Government of American Samoa
Pago Pago, American Samoa 96799
Overseas Operator
(Commercial Call (684) 663−4116)

ARIZONA

Arizona Department of Health
  Services
Office of Waste and Water Quality
2005 North Central Avenue
  Room 304
Phoenix, Arizona 85004
Hazardous Waste Management:
  (602) 255−2211

ARKANSAS

Department of Pollution Control
   and Ecology
Hazardous Waste Division
P.O. Box 9583
8001 National Drive
Little Rock, Arkansas 72219
(501) 562−7444

CALIFORNIA

Department of Health Services
Toxic Subtances Control Division
714 P Street, Room 1253
Sacramento, California 95814
(916) 324−1826
State Water Resources Control
   Board
Division of Water Quality
P.O. Box 100
Sacramento, California 95801
(916) 322−2867

COLORADO

Colorado Department of Health
Waste Management Division
4210 East 11th Avenue
Denver, Colorado 80220
(303) 320−8333 Ext. 4364

CONNECTICUT

Department of Environmental
   Protection
Hazardous Waste Management
   Section
State Office Building
165 Capitol Avenue
Hartford, Connecticut 06106
(203) 566−8843, 8844
Connecticut Resource Recovery
   Authority

179 Allyn Street, Suite 603
Professional Building
Hartford, Connecticut 06103
(203) 549−6390

DELAWARE

Department of Natural Resources
   and Environmental Control
Waste Management Section
P.O. Box 1401
Dover, Delaware 19903
(302) 736−4781

DISTRICT OF COLUMBIA

Department of Consumer and
   Regulatory Affairs
Pesticides and Hazardous Waste
   Materials Division
Room 114
5010 Overlook Avenue, S.W.
Washington, D.C. 20032
(202) 767−8414

FLORIDA

Department of Environmental
   Regulation
Solid and Hazardous Waste
   Section
Twin Towers Office Building
2600 Blair Stone Road
Tallahassee, Florida 32301
RE: SQG's
(904) 488−0300

GEORGIA

Georgia Environmental Protection
   Division
Hazardous Waste Management
   Program
Land Protection Branch

Floyd Towers East, Suite 1154
205 Butler Street, S. E.
Atlanta, Georgia 30334
(404) 656–2833
Toll Free: (800) 334–2373

GUAM

Guam Environmental Protection
  Agency
P.O. Box 2999
Agana, Guam 96910
Overseas Operator
(Commercial Call (671) 646–7579)

HAWAII

Department of Health
Environmental Health Division
P.O. Box 3378
Honolulu, Hawaii 96801
(808) 548–4383

IDAHO

Department of Health and Welfare
Bureau of Hazardous Materials
450 West State Street
Boise, Idaho 83720
(208) 334–5879

ILLINOIS

Environmental Protection Agency
Division of Land Pollution Control
2200 Churchill Road, #24
Springfield, Illinois 62706
(217) 782–6761

INDIANA

Department of Environmental
  Management

Office of Solid and Hazardous
  Waste
105 South Meridian
Indianapolis, Indiana 46225
(317) 232–4535

IOWA

U.S. EPA Region VII
Hazardous Materials Branch
726 Minnesota Avenue
Kansas City, Kansas 66101
(913) 236–2888
Iowa RCRA Toll Free:
  (800) 223–0425

KANSAS

Department of Health and
  Environment
Bureau of Waste Management
Forbes Field, Building 321
Topeka, Kansas 66620
(913) 862–9360 Ext. 292

KENTUCKY

Natural Resources and
  Environmental Protection
  Cabinet
Division of Waste Management
18 Reilly Road
Frankfort, Kentucky 40601
(502) 564–6716

LOUISIANA

Department of Environmental
  Quality
Hazardous Waste Division
P.O. Box 44307
Baton Rouge, Louisiana 70804
(504) 342–1227

MAINE

Department of Environmental
Protection
Bureau of Oil and Hazardous
Materials Control
State House Station #17
Augusta, Maine 04333
(207) 289–2651

MARYLAND

Department of Health and Mental
Hygiene
Maryland Waste Management
Administration
Office of Environmental Programs
201 West Preston Street, Room A3
Baltimore, Maryland 21201
(301) 225–5709

MASSACHUSETTS

Department of Environmental
Protection
Division of Solid and Hazardous
Waste
One Winter Street, 5th Floor
Boston, Massachusetts 02108
(617) 292–5589
(617) 292–5851

MICHIGAN

Michigan Department of Natural
Resources
Hazardous Waste Division
Waste Evaluation Unit
Box 30028
Lansing, Michigan 48909
(517) 373–2730

MINNESOTA

Pollution Control Agency

Solid And Hazardous Waste
Division
1935 West County Road, B-2
Roseville, Minnesota 55113
(612) 296–7282

MISSISSIPPI

Department of Natural Resources
Division of Solid and Hazardous
Waste Management
P.O. Box 10385
Jackson, Mississippi 39209
(601) 961–5062

MISSOURI

Department of Natural Resources
Waste Management Program
P.O. Box 176
Jefferson City, Missouri 65102
(314) 751–3176
Missouri Hotline:
(800) 334–6946

MONTANA

Department of Health and
Environmental Sciences
Solid and Hazardous Waste
Bureau
Cogswell Building, Room B-201
Helena, Montana 59620
(406) 444–2821

NEBRASKA

Department of Environmental
Control
Hazardous Waste Management
Section
P.O. Box 94877
State House Station
Lincoln, Nebraska 68509
(402) 471–2186

NEVADA

Division of Environmental
  Protection
Waste Management Program
Capitol Complex
Carson City, Nevada 89710
(702) 885–4670

NEW HAMPSHIRE

Department of Health and Human
  Services
Division of Public Health Services
Office of Waste Management
Health and Welfare Building
Hazen Drive
Concord, New Hampshire
  03301–6527
(603) 271–4608

NEW JERSEY

Department of Environmental
  Protection
Division of Waste Management
32 East Hanover Street, CN-028
Trenton, New Jersey 08625
Hazardous Waste Advisement
  Program: (609) 292–8341

NEW MEXICO

Environmental Improvement
  Division
Ground Water and Hazardous
  Waste Bureau
Hazardous Waste Section
P.O. Box 968
Santa Fe, New Mexico 87504–0968
(505) 827–2922

NEW YORK

Department of Environmental
  Conservation
Bureau of Hazardous Waste
  Operations
50 Wolf Road, Room 209
Albany, New York 12233
(518) 457–0530
SQG Hotline: (800) 631–0666

NORTH CAROLINA

Department of Human Resources
Solid and Hazardous Waste
  Management Branch
P.O. Box 2091
Raleigh, North Carolina 27602
(919) 733–2178

NORTH DAKOTA

Department of Health
Division of Hazardous Waste
  Management and Special Studies
1200 Missouri Avenue
Bismarck, North Dakota
  58502–5520
(701) 224–2366

NORTHERN MARIANA ISLANDS, COMMONWEALTH OF

Department of Environmental and
  Health Services
Division of Environmental Quality
P.O. Box 1304
Saipan, Commonwealth of
  Mariana Islands 96950
Overseas call (670) 234–6984

OHIO

Ohio EPA
Division of Solid and Hazardous
   Waste Management
361 East Broad Street
Columbus, Ohio 43266—0558
(614) 466—7220

OKLAHOMA

Waste Management Service
Oklahoma State Department of
   Health
P.O. Box 53551
Oklahoma City, Oklahoma 73152
(405) 271—5338

OREGON

Hazardous and Solid Waste
   Division
P.O. Box 1760
Portland, Oregon 97207
(503) 229—6534
Toll Free: (800) 452—4011

PENNSYLVANIA

Bureau of Waste Management
Division of Compliance Monitoring
P.O. Box 2063
Harrisburg, Pennsylvania 17120
(717) 787—6239

PUERTO RICO

Environmental Quality Board
P.O. Box 11488
Santurce, Puerto Rico 00910—1488
(809) 723—8184
   or
EPA Region II
Air and Waste Management
   Division

26 Federal Plaza
New York, New York 10278
(212) 264—5175

RHODE ISLAND

Department of Environmental
   Management
Division of Air and Hazardous
   Materials
Room 204, Cannon Building
75 Davis Street
Providence, Rhode Island 02908
(401) 277—2797

SOUTH CAROLINA

Department of Health and
   Environmental Control
Bureau of Solid and Hazardous
   Waste Management
2600 Bull Street
Columbia, South Carolina 29201
(803) 734—5200

SOUTH DAKOTA

Department of Water and Natural
   Resources
Office of Air Quality and Solid
   Waste
Foss Building, Room 217
Pierre, South Dakota 57501
(605) 773—3153

TENNESSEE

Division of Solid Waste
   Management
Tennessee Department of Public
   Health
701 Broadway
Nashville, Tennessee 37219—5403
(615) 741—3424

TEXAS

Texas Water Commission
Hazardous and Solid Waste
  Division
Attn: Program Support Section
1700 North Congress
Austin, Texas 78711
(512) 463–7761

UTAH

Department of Health
Bureau of Solid and Hazardous
  Waste Management
P.O. Box 16700
Salt Lake City, Utah 84116–0700
(801) 538–6170

VERMONT

Agency of Environmental
  Conservation
103 South Main Street
Waterbury, Vermont 05676
(802) 244–8702

VIRGIN ISLANDS

Department of Conservation and
  Cultural Affairs
P.O. Box 4399
Charlotte Amalie, St. Thomas
Virgin Islands 00801
(809) 774–3320
       or
EPA Region II
Air and Waste Management
  Division
26 Federal Plaza
New York, New York 10278
(212) 264–5175

VIRGINIA

Department of Health
Division of Solid and Hazardous
  Waste Management
Monroe Building, 11th Floor
101 North 14th Street
Richmond, Virginia 23219
(804) 225–2667
Hazardous Waste Hotline:
(800) 552–2075

WASHINGTON

Department of Ecology
Solid and Hazardous Waste
  Program
Mail Stop PV-11
Olympia, Washington 98504–8711
(206) 459–6322
In-State: 1–800–633–7585

WEST VIRGINIA

Division of Water Resources
Solid and Hazardous Waste/
  Ground Water Branch
1201 Greenbrier Street
Charleston, West Virginia 25311

WISCONSIN

Department of Natural Resources
Bureau of Solid Waste
  Management
P.O. Box 7921
Madison, Wisconsin 53707
(608) 266–1327

WYOMING

Department of Environmental
  Quality
Solid Waste Management Program
122 West 25th Street
Cheyenne, Wyoming 82002
(307) 777–7752
        or
EPA Region VIII
Waste Management Division
  (8HWM-ON)
One Denver Place
999 18th Street
Suite 1300
Denver, Colorado 80202–2413
(303) 293–1502

CHAPTER 25

REGULATIONS

I INTRODUCTION

To protect the employees, the community, and the environment, federal, state, and local authorities have established regulations covering the transportation, storage, handling, and disposal of hazardous materials. While this chapter describes some of the regulations that apply to the laboratory, its objective is not to provide a comprehensive source of information for any or all regulations. Moreover, the regulations included in this chapter do not represent all applicable regulations, nor the actual rulings themselves. Instead they are examples of the information contained within each regulation or act. In each case, the chapter cites the official title of each act or regulation. Laboratory staff should consult the original documents whenever in doubt.

II PRINCIPAL APPLICABLE REGULATIONS

All laboratory work should conform to the applicable local, state, and federal statutes that relate to occupational health and safety, transportation and handling, and environmental protection. Regulations of the Occupational Safety and Health Administration (OSHA) of primary concern to laboratories include the following portions of Title 29 of the Code of Federal Regulations (29 CFR):

- OSHA regulation 29 CFR, Part 1910, Section 134, (29 CFR 1910, 134), which is necessary for the establishment of a proper respiratory program, and
- OSHA regulation 29 CFR 1910. 1003–1910. 1016, which sets specific hood face velocities for specific substances.
- OSHA regulation 29 CFR 1910.1450, which addresses exposure to hazardous chemicals in laboratories.

Other important regulations are contained in Titles 10, 21, 29, and 49:

10 CFR: Regulation of Radioactive Materials

10 CFR 19 Notices, instructions, and reports to workers; inspections

10 CFR 20 Standards for protection against radiation

10 CFR 30 Rules of general applicability to domestic licensing of by-product material

10 CFR 31 General domestic licenses for by-product material

10 CFR 71 Packaging and transportation of radioactive material

10 CFR 170 Fees for facilities and materials licenses and other regulatory services under the Atomic Energy Act of 1954, as amended

21 CFR: Regulation of Controlled Substances

21 CFR 130 Federal requirements for controlled substances

29 CFR: Regulation of Workplace Hazards

29 CFR 1910, General industry standards

1910.20 Access to employee exposure and medical records

1910.133 Standard for eye and face protection

1910.134 Respiratory protection

1910.1000 Air contaminants

1910.1001–1910.1048 Employee exposures to Chemicals

1910.1200 Hazard communication

49 CFR: Regulation of Transportation

49 CFR 171 General information, regulations, and definitions

49 CFR 172.403 Hazardous materials tables and hazardous materials, (436, 438, 440) and communication regulations (556)

49 CFR 173 Subpart I General requirements for shipments and pack-

aging: Radioactive material

49 CFR 177.825 Carriage by public highway: Rating and training requirements for radioactive material

49 CFR 261–265 Waste management regulations

III OTHER APPLICABLE OSHA REGULATIONS (29 CFR 1910)

Part 1910 of Title 29 sets forth the minimum standards that employers are required to follow to protect their employees. This section first lists the CFR chapters that apply to laboratories, then highlights a few specific sections:

A. General
B. Adoption and Extension of Established Federal Standards
C. General Safety and Health Provisions
D. Walking and Working Surfaces
E. Means of Egress
F. Powered Platforms, Manlifts, Vehicle-Mounted Work Platforms
G. Occupational Health and Environmental Control
H. Hazardous Materials
I. Personal Protective Equipment
    1910.133 Eye and face protection
    1910.134 Respiratory protection
    1910.135 Occupational head protection
    1910.136 Occupational foot protection
    1910.137 Electrical protection devices
    1910.138 Effective dates
    1910.139 Sources of standards
    1910.140 Standards organizations
J. General Environmental Controls
K. Medical and First Aid
L. Fire Protection
M. Compressed Air and Gas Equipment
N. Materials Handling and Storage
O. Machinery and Machine Guarding
P. Hand and Portable Powered Tools
Q. Welding, Cutting, and Brazing
R. Special Industries
S. Electrical
T. Commercial Diving Operations
U–Y. (Reserved)
Z. Toxic and Hazardous Substances

A Eye and Face Protection (29 CFR 1910.133)

This regulation states that protective equipment for eyes and face is required where there is a reasonable probability that injury can be prevented by using such equipment. Each employer must make the equipment available. Hazards this regulation cites include flying objects, glare, liquids, injurious radiation, or any combination of them.

B Respiratory Protection (29 CFR 1910.134)

This section of the OSHA industrial hygiene regulations deals exclusively with respiratory protection and the requirements of respiratory protection programs. Specifics addressed in this section include but are not limited to selection, use, maintenance, and care of respirators; identification of gas mask canisters; and requirements for a minimal acceptable respiratory program. This section also defines the quality of the air or oxygen to be used and discusses procedures to be followed in a toxic or oxygen-deficient atmosphere.

C Air Contaminants (29 CFR 1910.1000)

This regulation is the core of the OSHA industrial hygiene requirements and provides permissible exposure limits (PELs) for approximately 400 chemicals. A PEL is the time-weighted average of airborne gases, vapors, and/or particulates to which an employee can be exposed. The PELs are expressed as parts of a chemical per million parts of air, and as milligrams of a chemical per cubic meter of air. OSHA specifically regulates these contaminants to be at or below the PEL during the employee's workshift. The regulation further defines the PELs and also short-term exposure limits.

D Employee Exposures to Chemicals (29 CFR 1910.1001– 1910.1043)

In addition to the materials cited in 29 CFR 1910.1000, approximately 30 chemicals are the subject of specific detailed regulations to control employee exposures. These regulations discuss PELs, engineering controls, personnel protective equipment, training, medical surveillance, and labeling.

E   Hazard Communication (29 CFR 1910.1200)

This federal regulation, implemented by OSHA in May 1986 and August 1987, supersedes most portions of state right-to-know laws (this should be reviewed for each state). The regulation focuses on employee notification of the hazards associated with the chemicals they are handling in their routine workday. The basic components of this regulation are employee training, labeling of hazardous chemicals, and procurement of material safety data sheets.

F   Occupational Exposure to Hazardous Chemicals in Laboratories (29 CFR 1910.1450)

This standard, issued by OSHA on January 31, 1990, requires each laboratory to have a chemical hygiene plan that includes standard operating procedures, criteria for handling and controlling toxic substances, and provisions for preventive maintenance and training.

G   Hazardous Waste Operations and Emergency Response (29 CFR 1910.120)

This OSHA standard regulates personnel protection against hazardous waste and emergency response activities. It may apply if a hazardous/ emergency response team exists at a given facility. The regulation cites requirements for training, protective equipment, and medical surveillance. This OSHA standard should be reviewed for its applicability to a laboratory's operation.

H   Access to Exposure and Medical Records (29 CFR 1910.20)

This standard requires employers to make exposure and medical surveillance records available to employees. It also requires retention of these records for at least 30 years.

IV   STATE RIGHT-TO-KNOW LEGISLATION

Before promulgation of the federal hazard communication regulation, many states and cities had developed their own right-to-know (RTK) laws. Similar to the federal act, these laws require that employees be informed of the dangers inherent in the hazardous chemicals they handle. However, the RTK laws in some states have components that require industry to notify state committees of the hazardous chemicals utilized at their worksites.

For this reason, the status of each state's RTK law should be reviewed to evaluate its impact on the laboratory.

V RESOURCE CONSERVATION AND RECOVERY ACT (40 CFR 261)

The Resource Conservation and Recovery Act of 1976 addresses all aspects of the waste management cycle, including waste generation, storage, transportation, and disposal. This act is administered by the U.S. Environmental Protection Agency (EPA). Any firm that generates, transports, treats, stores, or disposes of the specific hazardous wastes listed under Section 3001 of this act must notify the EPA within 90 days.

Title 40 (Parts 261–263 and 265.316) is pertinent to laboratories with small quantities of waste. Larger laboratories should refer to Parts 260–265.

VI STATE REGULATIONS

Individual states may enact specific regulations regarding chemical emission into air or water. All laboratories should acquaint themselves with state regulations of this type, which may apply to them. California's Proposition 65 is an example of a stringent state law that governs waste disposal. Proposition 65, also known as the Safe Drinking Water and Toxic Enforcement Act of 1986, has two major provisions: discharge prohibition and exposure warnings.

Proposition 65 prohibits chemicals that are known to the state of California to cause cancer or reproductive toxicity from being discharged into any drinking water sources. The ban becomes effective 20 months after being incorporated in a list of chemicals published by the governor. This control includes releases to water, land, or air, where the chemical would probably get into the drinking water. Proposition 65 also requires that anyone potentially exposed to a chemical on the list be adequately warned — for example, by labels, mailings, notices, and/or use of other public media.

VII EMERGENCY PLANNING AND COMMUNITY RIGHT-TO-KNOW PROGRAMS (40 CFR 300)

Under the Superfund Amendments and Reauthorization Act of 1986

(SARA), Congress included a provision (40 CFR 300) to require companies to report quantities of selected chemicals to local emergency response organizations. The purposes of the regulation are to encourage emergency planning efforts locally and to keep the community informed with respect to hazardous chemicals locally. Laboratories must be familiar with this regulation and must review chemical quantities to determine their obligations under the law.

VIII RADIOACTIVE USE AND TRANSPORTATION REGULATIONS (TITLES 10 AND 49)

Regulations pertaining to the use, disposal, and transportation of radioactive material can be found in the Code of Federal Regulations under Titles 10 and 49.

Title 10, written by the Nuclear Regulatory Commission (NRC), pertains to the licensing of radioactive material. It provides standards for protection against radiation, informs radiation workers of their rights, and describes the testing a package must undergo to be approved for radioactive material transport. It also covers several other areas of radioactive material usage that are not of concern to a laboratory.

The most pertinent portions of Title 10 are Parts 19 and 20. Instructions to workers (10 CFR 19) describes the rights of the radiation workers. The standards for protection against radiation (10 CFR 20) describe the rules a licensee must abide by.

In some states Title 10 does not apply to licensees. Several states have developed an agreement with the NRC under which the responsibilities of handling all aspects of managing and licensing radioactive materials in the state, except for commercial nuclear reactors, are delegated to the NRC. These states are known as "agreement states." It is important that a laboratory knows whether its state is an agreement state, since licensing will be handled accordingly. Frequently, agreement states simply adopt all NRC regulations as well as heeding their own concerns.

Title 49 concerns the labeling, packaging, transportation of, and route selection process for, radioactive materials. Part 172 of 49 CFR describes the labeling and placarding for radioactive packages and vehicles, respectively, and Part 177 discusses the routing requirements for vehicles carrying radioactive material.

The most important part of federal law relating to radioactive material is 10 CFR 20 (See Fig. 25–1).

PART 20–STANDARDS FOR PROTECTION AGAINST RADIATION

FIG. 25–1: Contents page of Part 20 of Title 10 of the Code of Federal Regulations.

IX CLEAN AIR ACT

The Clean Air Act led to the establishment of air quality standards in 1977. Section 111 of this act requires the EPA to publish a list of substances that present potential human health hazards and to define specific control requirements. Section 112 requires the EPA to generate a list of substances that may create chronic health problems in humans. The act also instituted ambient air quality standards, known as National Ambient Air Quality Standards, which govern the amount of a material that can be released into the atmosphere. State regulations should also be consulted (see Section VI).

X CLEAN WATER ACT

The Clean Water Act of 1977 requires the EPA to list chemicals or materials that are considered to be hazardous when spilled in waterways. The act also requires the EPA to establish ambient water quality criteria that dictate the amounts of certain materials that may be released into navigable waters. The act also defines spill cleanup procedures and institutes penalties in the event of an accidental spill. State regulations should also be consulted (see Section VI).

XI TOXIC SUBSTANCES CONTROL ACT (TSCA)

The Toxic Substances Control Act (TSCA) of 1976 is primarily concerned with the EPA's list of chemicals that require toxicity testing. The EPA selected these chemicals on the basis of quantity to be manufactured, quantity entering the environment, number of individuals exposed, and expected extent of exposure. This act also defined the rights of the EPA to request the chemical industry to submit existing safety and health studies for certain substances.

XII INFORMATION SOURCES

The acts and regulations discussed in this chapter represent only federal (and some state) rulings that may apply to toxicology laboratories. Each laboratory also falls under the jurisdiction of its state and local governmental regulations.

Copies of these requirements, in their entirety, may be obtained from the Government Printing Office, the National Technical Information Service, or the specific state and local offices responsible for such information.

BIBLIOGRAPHY

Alden, J.L., and Kane, J.M., *Design of Industrial Ventilation Systems*, 5th ed., Industrial Press, New York, 1982.

American Conference of Governmental Industrial Hygienists, *Industrial Ventilation, A Manual of Recommended Practice*, latest bi-ennial edition, Lansing, MI, 1986.

ACGHI, *Documentation of TLV's and BEI's*, 5th ed, Cincinnati, OH, 1986.

AIHA Biohazards Committee, *Biohazards Reference Manual*, American Industrial Hygiene Association, 1985.

American Iron and Steel Institute, *Fire Protection through Modern Building Codes*, Washington, D.C., 1971.

American National Standard for Emergency Eyewash and Shower Equipment, ANSI 2358.1, American National Standards Institute, New York, 1981.

American National Standard for Men's Safety—Toe Footwear, ANSI 241, American National Standards Institute, New York, 1983.

American National Standard for Occupational and Educational Eye and Face Protection, ANSI 787.1, American National Standards Institute, New York, 1979.

American National Standard Practice for Industrial Lighting, ANSI Standard RP-7, American National Standards Institute, New York, 1979.

American National Standard for Protective Headwear for Industrial Workers, ANSI 789.1, American National Standards Institute, New York, 1981.

Baselt, R.D., *Biological Monitoring Methods for Industrial Chemicals*, Biomedical Publications, Davis, CA, 1980.

Biotechnology, Inc., "OSHA Medical Surveillance Requirements and NIOSH Recommendations," prepared for the National Aeronautics and Space Administration, January 1980.

Boyce, P.R., *Human Factors in Lighting*, MacMillan, New York, 1981.

Braker, W., and Mossman, A.L., *Matheson Gas Data Book*, 5th ed., Matheson Gas Products, East Rutherford, NJ, 1971.

Braker, W., Mossman, A.L., and Siegel, D., Eds., *Effects of Exposure to Toxic*

Gases — First Aid and Medical Treatment, 2nd ed., Matheson, Lyndhurst, NJ, 1977.

Brannigan, F.L., *Building Construction for the Fire Service*, National Fire Protection Association, Boston, MA, 1971.

Bretherick, L., *Handbook of Reactive Chemical Hazards*, 3rd ed., Butterworth, Boston, MA, 1985.

Bretherick, L., *Hazards in the Chemical Laboratory*, 3rd ed., The Royal Society of Chemistry, Burlington House, London, 1981.

Bryan, J.L., and Picard, R.C., Eds., *Managing Fire Services*, International City Management Association, Washington, D.C., 1979.

Cakir, A., Hart, D.J., and Stewart, T.F.M., *Visual Display Terminals*, Wiley, New York, 1980.

Caplan, K.J., and Knutson, G.W., "The Effect of Room Air Challenge on the Efficiency of Laboratory Fume Hoods," *ASHRAE Transactions*, Vol. 83, Part I, 1977.

Castegnaro, M., and Sansone, E.B., Eds., *Chemical Carcinogens: Some Guidelines for Handling and Disposal in the Laboratory*, Springer-Verlag, New York, 1986.

Chamberlin, R.I., and Leahy, J.E., "A Study of Laboratory Fume Hoods," U.S. EPA report, prepared under contract No. 68-01-4661, 1978.

Chatigny, M.A., and Clinger, D.I., "Contamination Control in Aerobiology," in *An Introduction to Experimental Aerobiology*, Dimmick, L., and Akers, A.B., Eds., Wiley, New York, 1969.

Clark, W.E., *Firefighting Principles and Practice*, Donnell Publishing, New York, 1974.

Clayton, G.D., and E.E. Clayton, Eds., *Patty's Industrial Hygiene and Toxicology*, 3rd revised edition, Vols. 1, 2A, 2B, 2C, 3A, 3B, Wiley, New York, 1978.

Code of Federal Regulations, *Food and Drugs*, No. 21, Part 1300 to End, 1983.

Collins, C.H., *Laboratory-acquired Infections*, Butterworth, Boston, MA, 1983.

Damon, A., Stoudt, H.W., and McFarland, R.A., *The Human Body in Equipment Design*, Harvard University Press, Cambridge, MA, 1966.

Dangerous Goods Regulations, 28th ed., IATA, 1987.

Dreisbach, R.H., *Handbook of POISONING: Prevention Diagnosis Treatment*, 10th ed., Lange Medical Publications, Los Altos, CA, 1983.

Everett, K., and Hughes, D., *A Guide to Laboratory Design*, Butterworth, London, 1981.

Factory Mutual Engineering Corporation, *Handbook of Industrial Loss Prevention*, 2nd ed., McGraw-Hill, New York, 1967.

Fuller, F.H., and Etchells, A.W., "The Rating of Laboratory Hood Performance," *ASHRAE Journal*, October 1979.

Fuscaldo, A.A., Erlich, B.J., and Hindman, B., Eds., *Laboratory Safety — Theory and Practice*, Academic Press, New York, 1980.

Gaffney, L.F., et al., "Field Testing and Performance Certification of Laboratory Fume Hoods," presented at Industrial Hygiene Conference, May, 1980.

Handling Chemicals Safely, 2nd ed., Dutch Association of Safety Experts, 1980.

Harless, J., "Components in the Design of a Hazardous Chemicals Handling Facility," in *Health and Safety for Toxicity Testing*, Butterworth, Boston, MA, 1984.

Harrington, J.M., and Gill, F.S., Eds., *Occupational Health*, Blackwell Scientific Publications, Oxford, London.

Hatayama, H.K., et al., EPA Document 600/2-80-076, *A Method for Determining the Compatibility of Hazardous Waste*, U.S. Government Printing Office, Washington, D.C., 1980.

Health and Safety in the Chemical Laboratory—Where do we go from here?, No. 51, The Royal Society of Chemistry, Burlington House, London, 1985.

Ho, M.H., and Dillon, H.K., *Biological Monitoring of Exposure to Chemicals*, Wiley, New York, 1987.

Holt, G.L., *Employee Facial Hair versus Employer Respirator Policies*, Appl. Ind. Hyg. J., Vol. 2, No. 5, September 1987.

Hoover, B.K., Baldwin, J.K., Velnar, A.F., Whitmire, C.E., Davies, C.L, and Bristol, D.W., *Managing Conduct and Data Quality of Toxicology Studies*, Princeton Scientific Publishing, Princeton, NJ, 1986.

Huchingson, R.D., *New Horizons for Human Factors in Design*, McGraw-Hill, New York, 1981.

Jackson, H.S., McCormack, W.B., Rondestvedt, G.S., Smeltz, K.C., and Viele, I.E., "Safety in the Chemical Laboratory," J. Chem. Educ., Vol. 47, No. 3, A176, March 1970.

Jonathan, R.D., "Selection and Use of Eyewash Fountains and Emergency Showers," Chem. Eng., pp. 147–150, Sept. 15, 1975.

Keith, L.H., and Walters, D.B., *Compendium of Safety Data Sheets for Research and Industrial Chemicals*, Parts I–VI, Sections 1–3, VCH Publishers, 1985.

Keith, L.H., and Walters, D.B., *Compendium of Safety Data Sheets for Research and Industrial Chemicals*, Parts I–VI, Sections 4–6, VCH Publishers, 1987.

Konz, S., *Work Design*, Grid Publishing, Columbus, OH, 1979.

Kuehne, R.W., "Biology Containment Facility for Studying Infectious Disease," Appl. Microbiol., Vol. 26, 1973, pp. 239–245.

Laboratory Decontamination and Destruction of Carcinogens in Laboratory Wastes: Some Polycyclic Aromatic Hydrocarbons, IARC Scientific Publications, No. 49.

Leong, B.K.J., *Proceedings of the Inhalation Toxicology and Technology Symposium*, Ann Arbor Science Publishers, Ann Arbor, MI, 1981.

Levy, B.S., and Wegman, D.H., Eds., Occupational Health: Recognizing and Preventing Work-related Disease, Little, Brown, Boston, MA, 1983.

Linch, A.L., *Biological Monitoring for Industrial Chemical Exposure Control*, CRC Press, West Palm Beach, FL, 1974.

Martin and Harbison, *An Introduction to Radiation Protection*, 2nd ed., Chapman and Hall Ltd., 1979.

Martin, W.F., Lippitt, J.M., and Prothero, T.G., Eds., *Hazardous Waste Handbook for Health and Safety*, Butterworth, Boston, MA, 1987.

Material Safety Data Sheets, Genium Publishing, Schenectady, NY (updated periodically).

McCormick, E.J., *Human Factors in Engineering and Design*, 4th ed., McGraw-Hill, New York, 1976.

McDermontt, H.J., *Handbook of Ventilation for Contaminant Control*, 2nd ed., Butterworth, Boston, MA, 1985.

McKinnon, G.P., Ed., *Fire Protection Handbook*, 14th ed., National Fire Protection Association, Quincy, MA, 1976.

Messinger, H.B., Clappo, R., Nolan, P., and Stagner, L., "An Analysis of Medical Monitoring Data Required by OSHA Health Regulations," U.S. Department of Labor, Report No. ASPER/CON-78/0167/A, 1979.

Miller, C.D., Songer, J.R., and Sullivan, J.F., "A Twenty-five Year Review of Laboratory-acquired Human Infections at the National Animal Disease Center," Amer. Ind. Hyg. Ass. J., Vol. 48, No. 3, pp. 271–275, 1987.

Mond. C., et al., "Human Factors in Chemical Containment Laboratory Design," Amer. Ind. Hyg. Ass. J., Vol. 48, No. 10, pp. 823–830, 1987.

Muller, K.R., et al., *Chemical Waste Handling and Treatment*, Springer-Verlag, New York, 1986.

National Archives and Records Administration, Office of Federal Register, Code of Federal Regulations 29, Parts 1900 to 1910, Occupation Safety and Health Administration Institute, 1986.

National Archives and Records Service, Office of the Federal Register, General Services Administration, Code of Federal Regulations 21, Government Printing Office, 1986.

National Commission on Fire Prevention and Control, *America Burning: A Report of the National Commission on Fire Prevention and Control*, U.S. Government Printing Office, Washington, D.C., 1973.

National Fire Protection Association, *Fire Protection for Laboratories Using Chemicals, NFPA 45-1986*, Quincy, MA, 1982.

National Fire Protection Association, *Flammable Liquids Code, NFPA No. 30*, National Fire Codes, Quincy, MA, 1984.

National Fire Protection Association, *Life Safety Code, NFPA No. 101*, National Fire Codes, Quincy, MA, 1980.

National Fire Protection Association, *Life Safety Code, NFPA 101-1981*, National Fire Codes, Quincy, MA, 1985.

National Fire Protection Association, *National Electrical Code, NFPA 70-1987*, Quincy, MA, 1981.

National Institute for Occupational Safety and Health, "Work Practices Guide for Manual Lifting," NIOSH publication No. 81–122, Cincinnati, OH, 1981.

National Safety Council, *Accident Prevention Manual for Industrial Operations*, 7th ed., Chicago, IL, 1974.

National Sanitation Foundation, "National Sanitation Foundation Standard No. 49 for Class II (Laminar Flow) Biohazard Cabinetry," National Sanitation Foundation, Ann Arbor, MI, 1976.

O'Connor, C.H., and Lirtzman, S.I., *Handbook of Chemical Industry Labeling*, Noyes Publications, Park Ridge, NJ, 1984.

Performance of Protective Clothing, ASTM Special Technical Publication 900, Philadelphia, PA, 1986.

Phillips, G.B., and Runkle R.S., "Design of Facilities for Microbial Safety," in *CRC Handbook of Laboratory Safety*, 2nd ed., and Steere, N.V., *CRC Handbook of Laboratory Safety*, 2nd ed., CRC Press, West Palm Beach, FL, 1971.

Piotrowski, J.K., "Exposure Tests for Organic Compounds in Industrial Toxicology," National Institute for Occupational Safety and Health (NIOSH), DHEW (NIOSH) Publication No. 77−144, 1977.

Poulton, E.C., *Environment and Human Efficiency*, Charles C. Thomas, Springfield, IL, 1970.

Proctor, N.H., and Hughes, J.P., *Chemical Hazards of the Workplace*, J.B. Lippincott Company, Philadelphia, PA, 1978.

Prudent Practices for Disposal of Chemicals from Laboratories, National Research Council, Washington, D.C., 1983.

Prudent Practices for Handling Hazardous Chemicals in Laboratories, National Research Council, Washington, D.C., 1981.

Registry of Toxic Effects of Chemical Substances, 1983−84 Cumulative Supplement to the 1981−82 edition, DHHS (NIOSH) Publication No. 86−103.

Re Velle, J.B., *Safety Training Methods*, Wiley, New York, 1980.

Rothstein, M.A., *Medical Screening of Workers*, The Bureau of National Affairs, Washington, D.C., 1984.

Rom, W.N., *Environmental and Occupational Medicine*, Little, Brown, Boston, MA, 1983.

Runkel, R.S., and Phillips, G.B., *Microbial Containment Control Facilities*, Van Nostrand Reinhold, New York, 1969.

Russell, H., "Tempered Water for Safety Showers and Eye Baths," Chemical Engineering, Nov. 24, 1982.

Safe Storage of Laboratory Chemicals, Pepitone, D.A., ed., Wiley, New York, 1984.

Sansone, E.B., and Jonas, L.A., "The Effect of Exposure to Daylight and Dark Storage on Protective Clothing Material Permeability," Amer. Ind. Hyg. Ass. J., Vol. 42, 1981, pp. 841−843.

Sax, I.N., *Dangerous Properties of Industrial Materials*, 6th ed., Van Nostrand Reinhold, New York, 1984.

Schulle, H.E., "Personal Protective Devices," in *The Industrial Environment—Its*

Evaluation and Control, National Institute for Occupational Safety and Health, Cincinnati, OH, 1973.

Schwope, A.D., et al., *Guidelines for the Selection of Chemical Protective Equipment*, 2nd ed., American Conference of Governmental Industrial Hygienists, Cincinnati, OH, 1985.

Scientific Apparatus Makers Association, Standard LF10-1980, Laboratory Fume Hoods, Washington, D.C., 1980.

Scott, R.A., and Doemeny, L.J., Eds., *Design Considerations for Toxic Chemical and Explosive Facilities*, American Chemical Society, Washington, D.C., 1987.

Shleien, T., *The Health Physics and Radiological Health Handbook*, Neicleon Lectern Associates, 1984.

Sittig, M., *Handbook of Toxic and Hazardous Chemicals and Carcinogens*, 2nd ed., Noyes Publications, Park Ridge, NJ, 1985.

Slein, M.W., and Sansone, E.B., Eds., *Degradation of Chemical Carcinogens: An Annotated Bibliography*, Van Nostrand Reinhold, New York, 1980.

Slote, L., *Handbook of Occupational Safety and Health*, Wiley, New York, 1987.

Stuart, D.G., First, M.W., Jones Jr., R.L., and Eagleson Jr., J.M., "Comparison of Chemical Vapor Handling by Three Types of Class II Biological Safety Cabinets," Particulate and Microbial Controls, Vol. 2, 1983, pp. 18–24.

Suspect Chemicals Sourcebook: A Guide to Industrial Chemicals Covered under Major Federal Regulatory and Advisory Programs, 6th ed., Roytech Publications, 1987.

Tegeris, A.S., *Toxicology Laboratory Design and Management for the 80's and Beyond*, Karger, New York, 1984.

Toxic and Hazardous Industrial Chemicals Safety Manual for Handling and Disposal with Toxicity and Hazard Data, The International Technical Information Institute, Tokyo, Japan, 1978.

Tucker, M.E., Industrial Hygiene: A Guide to Technical Information Sources, Amer. Ind. Hyg. Ass. J., Akron, OH, 1984.

U.S. Coast Guard. *CHRIS Hazardous Chemical Data* — Commandant Instruction M.16465.12A., U.S. Government Printing Office 0-479-762:QL3, Washington, D.C., U.S. Department of Transportation, 1985.

U.S. Department of Commerce, National Bureau of Standards, *Safe Handling of Radioactive Materials*, Handbook, Vol. 92, 1964.

U.S. Department of Health, Education and Welfare, NIH Biohazards Safety Guide, 1974.

U.S. Department of Health, Education and Welfare, National Institute for Occupational Safety and Health, "Criteria for a Recommended Standard Occupational Exposure to Malathion," HEW Publication No. (NIOSH) 76–205, 1976.

U.S. Department of Health, Education and Welfare, "Design Criteria for Viral Oncology Research Facilities," DHEW Publication No. (NIH) 75–891, Washington, D.C., U.S. Government Printing Office, 1975.

U.S. Department of Health, Education and Welfare, Occupational Exposure Sampling Strategy Manual, DHEW (NIOSH) Publication No. 77—173, 1977.

U.S. Dept. of Health, Education and Welfare, Recommended Industrial Ventilation Guidelines, HEW Pub. No. 76—162, The National Institute of Occupational Safety and Health, Contract No. CDC-99-74-33, prepared by Arthur D. Little, Cambridge, MA., U.S. Government Printing Office, 1976-657/5543, January 1976.

U.S. Department of Health and Human Services, 1st ed., CDC/NIH Biosafety in Microbiological and Biomedical Laboratories, March 1984.

U.S. Department of Health and Human Services, National Institute of Health, "NIH Guidelines for the Laboratory Use of Chemical Carcinogens," NIH Publication No. 81—2385, May 1981.

U.S. Department of Health and Human Services, National Institute of Health *NIOSH Pocket Guide to Chemical Hazards*, U.S. Government Printing Office, Washington, D.C., September 1985.

U.S. Department of Labor, Occupational Safety and Health Administration, Subpart 2 — Toxic and Hazardous Substances, 1910.1000-1910-1045, *General Industry* (rev.), U.S. Government Printing, Office, Washington, D.C., 7 November 1978.

Van Cott, H.P., and Kinkade, R.G., Eds., Human Engineering Guide to Equipment Design (revised edition), Government Printing Office, Washington, D.C., 1972.

Verschueren, K., *Handbook of Environmental Data on Organic Chemicals*, 2nd ed., Van Nostrand Reinhold, New York, 1983.

Walls, E.L., Safety in the Chemical Laboratory: Laboratory HVAC Systems — A Troika, J. Chem. Educ., Vol. 64, No. 12, December 1987.

Walters, D.B., and Jameson, C.W., *Health and Safety for Toxicity Testing*, Butterworth, Boston, MA, 1984.

Walters, D.B., *Safe Handling of Chemical Carcinogens, Mutagens, Teratogens, and Highly Toxic Substances*, Vols. 1 and 2, Ann Arbor Science Publishers, Ann Arbor, MI, 1980.

Walters, D.B., Stricoff, R.S., and Ashley, L.E., *The Selection of Eyewash Stations for Laboratory Use*, (unpublished report).

Waritz, R.S., "Biological Indicators of Chemical Dosage and Burden," in *Patty's Industrial Hygiene and Toxicology*, Volume III, Cralley, L.J., and Cralley, L.V., Eds., Wiley, New York, 1979.

Williams, J.R., "Evaluation of Intact Gloves and Boots for Chemical Permeation," Amer. Ind. Hyg. Ass. J., Vol. 42, 1981, pp. 468—476.

Woodson, W.E., *Human Factors Design Handbook*, McGraw-Hill, New York, 1981.

Young, G.S., "Laboratory Worker Medical Surveillance" in *Health and Safety for Toxicity Testing*, Walters, D.B., and C.S. Jameson, Eds., Butterworth, Boston, MA, 1984.

INDEX